SHEDDING YOUR SALES SHARK

Lessons from a Recovering Sales Shark &
How to Apply Them to
Social Media Marketing & Life

by Shelley R. Roth

SPRINGBOARD PUBLISHING

SPRINGBOARD PUBLISHING

©2012 by Shelley R. Roth

www.springboardworks.com

Jacket design: Shelley Roth and Energy Arts Alliance
Editor: Energy Arts Alliance
Graphic Designer: Tric Ortiz
Cover photo by: Yemi Owolabi

LIBRARY OF CONGRESS CATALOGING-IN-PUBLICATION DATA
Roth, Shelley R.
SHEDDING YOUR SALES SHARK: *Lessons from a Recovering Sales Shark & How to Apply Them to Social Media Marketing & Life*
/ Shelley R. Roth. - 1st ed.
p. cm.
1. Marketing & Sales 2. Social media

ISBN-10: 098387042X | ISBN-13: 978-0-9838704-2-5

PRINTED IN THE UNITED STATES OF AMERICA

First Edition

Learn more at www.shelleyroth.com

CONTENTS

CHAPTER 3

INTRODUCTION

- Are you working in a corporate setting where creative thinking and feeling with the heart are frowned upon?

- Are you starting to question whether the sales and corporate strategies you were taught are ethical or obsolete?

- Do you dread going to work because you feel like you can't be yourself?

- Do you find yourself being a really different person at home with your family and friends compared to who you are at work?

- Have you ever found yourself sacrificing your principles to make a sale or get something done in business?

- Are you sick of having to make quota every month and think there must be a better way?

- Do you feel like you're getting left behind by social media and don't know where to start learning more about it?

- Are the sales and business tactics you always used successfully before suddenly not working so well?

- Do you know or manage someone in sales and want to give them new ways to be successful?

Then this book is for you.

This is the story of a recovering sales shark. It is a story about how I shed my sales shark persona and learned all about serving vs. selling or giving to grow vs. giving to get. It is the story of how one of the best salespeople in the world learned about the importance of getting real and focusing on building a community instead of building a nest egg. It is the story of how the social media revolution has sent the sales shark the way of the dinosaur and how to use today's top social media tools to set you and your business free.

This book will guide you through my story: The transition I made from corporate player to social media navigator and the universal lessons I learned along the way.

It will take you through my "dark side" when it was all about closing the deal…signing the order…selling more stuff. It takes you first-hand through the scarcity principal vs. the mindset of abundance.

The first chapter, "From Sales Shark to Recovering Sales Shark," shares seven lessons learned by getting real and offers some concrete tips to help you shed the sales shark skin – or help someone you care about or manage shed that skin.

The second chapter, "Why Sales Sharks Are Becoming Dinosaurs: The Evolution of the Market, Marketing and Sales," makes the case for why sales sharks have become obsolete and offers insights on alternative ways to build and connect to your community effectively by being real and transparent with your community.

The third chapter, "Social Soup: A Guide to Social Media," describes the major social media options available to you and provides easy ways to use them.

Throughout our journey together, I will shed light on how to use social networks to give to your community of Fans, Friends, Followers and Connections, all while living the philosophy of "give to grow."

My journey as a recovering sales shark has had many roadblocks and speed bumps along the way. However, upon reflection, every obstacle encountered, every "no" I heard, every place I lost my job ultimately brought me to this moment in time so I can live in a new way – a way that isn't usually rewarded in the sales world.

The most important lesson I learned in my evolution from sales shark to recovering sales shark is that if every company out there – be it a solopreneur operation or a *Fortune* 100 – came from a place of truly serving their community, adding value, giving to grow with a recognition that money and abundance flow naturally when you are doing right by others without any selfish motive, I believe our world would transform. Through my own personal journey, I have found that giving without thought to getting makes a huge difference. It makes a difference in who I am in the world and what I attract.

When we come from scarcity, we attract less. When we come from abundance....well, guess what you get?

Whether you are a sales shark who is starting to question "eat-what-you-kill," excessively quota-driven sales strategies, a recovering sales shark or a fully recovered sales shark, it is my hope that this book will inspire you to live and work with a new and maybe unexpected mindset – supported and made possible by social media.

CHAPTER 1
From Sales Shark to Recovering Sales Shark:
Lessons Learned by Getting Real

In this chapter, we're going to explore seven important universal lessons that I have learned on my journey from sales shark to recovering sales shark that I hope will help you on the road to setting you and your business free online and in life with a platform for creating abundance for yourself and your business.

But first, let's define what I mean by "sales shark" and set the stage of how the sales shark that was Shelley was born.

My definition of a "sales shark" isn't limited to those in sales. I use the term "sales shark" to describe anyone who is in a job that may be defining them, mostly in a corporate setting, where creative thinking and feeling with the heart are frowned upon.

Setting the Stage: A Sales Shark is Born

Here's my story of NOT being real for the first two-thirds of my life...Yup, two-thirds. Wow, what a waste of time and energy being someone I wasn't! Why was I not ME? I truly believed that the "real me" wasn't like "normal" people. I particularly didn't see the "real me" in "rich" and successful people. Both personally and professionally, I had on every mask I could use to be anything but ME! Why?

Let's start from the very beginning when I was labeled a "tomboy." I was ridiculed (and not just by school-aged friends but also by many adults in my life) for wanting to excel at sports, be competitive, win at everything I attempted. Competition was not "lady-like"...If I had a dollar for every time I heard that term, I could feed all of Africa!

I mean, back then, ladies and little girls were supposed to behave a certain way. Dainty? Not me! Quiet and unassuming? Not this one! Speak only when spoken to? Forget about it....Nope, I didn't fit the mold of whatever someone decided good little girls were supposed to be: sugar and spice and everything nice....Yeah, right!

The other thing I heard repeatedly during my childhood and teen years was, "You are a miserable soul!" In my experience, perception is reality, so hearing this over again, well, I naturally thought I was basically a miserable soul.

Now this is a book about business and shedding the sales shark and being real, so I will not digress too much to my childhood. I just want to set the stage for my great awakening: **the fact that being REAL is a gift.**

My life's purpose is to encourage others to be real by being real myself. I'm here to help remind people – especially sales professionals – that being REAL is what we all should strive for – no masks, no pretense, just authenticity and transparency with the by-products of joy and abundance.

So professionally, where did I start? Well, I was always a businessperson with a special gift for sales – from my early years with very successful lemonade stands, ice ball stands, magazine sales and selling cutlery door-to-door. It was all about excelling and winning. I watched my father sell, and he was very successful, winning trips to many first-class exotic destinations as a result of selling. (Actually, he was selling snow to Eskimos, and again, the learned behavior was that it was okay to do this.)

But my life as a sales shark was not to be, at least not yet. When I was young, "businessperson" just didn't fit the girls' career options when it was time to pick a college major. So I started off going the safe route, the route every girl opted for when I was graduating: nurse, secretary or teacher. Teacher was the one of the three that felt most appropriate. (I was always told I was a Pied Piper with children.) In retrospect, it was a great foundation for who I am today, albeit, most likely not the perfect route for this "tomboy" (or so I thought).

So, teacher was a safe place for a girl who felt like a boy. It was a noble career, and I totally enjoyed working with elementary-age students. I spent eight years in education as a teacher, guidance counselor and assistant principal – all enjoyable, all successful, but no apparent brass ring. I wanted "stuff," a lot of stuff. Stuff = success, right? So I thought I would follow in my dad's footsteps. I sold my house, quit my job, cut my hippie hair and moved on down with a U-Haul to Houston, Texas.

Yep, sales would bring me all the things I thought would satisfy this lost, "miserable" soul.

I found out that sales was a place for elephant hunters: "kill or be killed," "scorch the earth," "take no prisoners"... and I never felt that way. I was a farmer, a nurturer. I didn't want to eat what I killed. I didn't want to kill; I wanted to nurture. But that didn't matter. Part of me wanted those trips to Hawaii, those luxury cars, the big house. At least I thought I did, and that part of me won out over the nurturer.

My dad was motivated by stuff, I thought. (I never really got to ask him if that was really true.) I think the middle-class dream is what gets to many of us, and we think success is the stuff we acquire. We think it's a symbol of who we are and what we have accomplished. But, I digress...

So, I was rewarded for the things that I ultimately learned I truly didn't believe in, things I didn't think were really signs of success, things that never really brought me happiness or joy.

Before I was "discovered" by a top computer software company, I paid my dues during my first two years in Houston. From selling frozen food plans to people on welfare (mostly chicken because the bones weighed a lot) to sun-tanning salon franchises, I truly went through the school of hard-knocks sales. Before being discovered by that computer company, I sold its regional manager a swimming pool and, alas, I was plucked from obscurity into a top-notch company.

That first *Fortune* 500 I worked for had 80 salespeople, and I was the number one quota achiever in my first year. I "won" a two-week, all-expenses paid trip to Tahiti, and I had 79 male counterparts mad as hornets about that! Being the practical person I am, I decided to forgo the trip to Tahiti and instead took a first-class trip to Hawaii for a week, using the balance of the award to purchase many items that I had never owned before: patio furniture, huge TV, fancy lamps, etc. Wow, that was some first year with the big company! But what did I want? What were my targets, my monetary rewards? Cars, houses, boats??? Nope. When asked by the sales manager to list all of the "things" we aspired to and wanted (which was used to motivate us to sell more), peace, harmony, joy were at the top of my list...Talk about a fish out of water...But, boy, did I learn to play the role, to wear the mask, to create the avatar!

In addition to all the past sales experiences that I've already shared, there was something else that really taught me to be an excellent salesperson. It happened when I was five years old. It was around Halloween, and we were at a costume contest at the neighborhood community center. I was dressed in a homemade bunny costume, and I was really feeling it. Hip-hopping my little heart out. And then, I WON the contest. I was called up to the stage to accept my reward and walked right up bold as could be. The Master of Ceremonies was announcing this into his wired microphone as he handed little bunny Shelley a shiny silver dollar and...BAM!...The little bunny literally had the shock of her little life...The microphone wasn't grounded, the coin acted as a conductor and - WHAM! - an electric shock was transmitted from the microphone through the coin into my body. What was a bunny to do?

This was the moment the salesperson was born. Did I listen to my heart and cry and ask for help or comfort? Did I yell out for my mom to come up there and take me away? NOOO, I stood taller, stronger and pretended that there was nothing wrong. That was the moment I listened to my head, and it said, "Don't be a sissy. Don't cry. Don't show emotion. PRETEND there's nothing wrong. PRETEND you are happy. You don't want anyone to see your real feelings or heart... It's all about looking good!" From that day forward, I was terrified of public speaking, but oh, what a great actor I became!

After 20 years of being the great salesperson, working for *Fortune* 500 companies, listening to my managers tell me how to close more deals and "winning" trips to Hawaii (6 times), Bermuda, The Bahamas, Lake Tahoe, Cancun, Amsterdam, Belgium, China, buying the house, the fancy cars, all the stuff, well, I still didn't feel fulfilled. What was my purpose? Why was I on the Earth?

Those questions must have set something in motion. After all those years at the top, I was fired because I finally couldn't do what I was told since in my heart I knew that selling ice to Eskimos wasn't a noble deed.

After being kicked out of the corporate nest, I started my own business and have worked with companies over the last decade on what I know - growing sales. A couple of years back, I was intrigued by social media, so, sponge that I am for learning, I decided to create curricula on various social media tools and combine my passion for business with my love of training, and here I am today. The more I worked in social media, the more I realized that we all need to be aware of what is REAL out there. That's when I was led by my wonderful energy advisors to discover my reason for being on the Earth and to encourage others to do the same: *to be REAL.* And this was the one thing I never had confidence in being. This was the one thing I thought was BAD. Every time I just WAS from my heart, it got slapped down... But not anymore.

For many of us, getting real seems risky. With all the criticisms most of us experience over the years - all the times we were told to be quiet, to follow the rules, to be "lady-like" or "man up," to get a real job - sometimes we've buried our real selves so deeply in trying to meet others' expectations or gain acceptance, we don't even know who our real self is. But the classic children's story, *The Velveteen Rabbit*, describes the whole truth of being real in one scene:

"What is REAL?" asked the Rabbit one day, when they were lying side by side near the nursery fender, before Nana came to tidy the room. "Does it mean having things that buzz inside you and a stick-out handle?"

"Real isn't how you are made," said the Skin Horse. "It's a thing that happens to you. When a child loves you for a long, long time, not just to play with, but REALLY loves you, then you become Real."

"Does it hurt?" asked the Rabbit.

"Sometimes," said the Skin Horse, for he was always truthful. "When you are Real you don't mind being hurt."

"Does it happen all at once, like being wound up," he asked, "or bit by bit?"

"It doesn't happen all at once," said the Skin Horse. "You become. It takes a long time. That's why it doesn't happen often to people who break easily, or have sharp edges, or who have to be carefully kept. Generally, by the time you are Real, most of your hair has been loved off, and your eyes drop out and you get loose in the joints and very shabby. But these things don't matter at all, because once you are Real you can't be ugly, except to people who don't understand."

I can tell you that REAL is good. REAL is king. REAL is what we all should strive for. REAL is the gift I was given.

So now, I declare my vision and purpose: To help and encourage people – especially recovering sales sharks – to be REAL. To listen to your HEART. It knows what's right.

More of the Story: Software XYZ

From the minute I began the intensive interview process at the *Fortune 500* software company, the experience never felt right to me in my heart. It felt great for my ego; I mean, who wouldn't want to work at Software XYZ? People were and still are clamoring to work there. They have onsite workout facilities, a 4-star restaurant, hair and nail salon, banking and basketball court – all on the premises! This was the bait on the hook along with a fantastic, beautiful first-class building with top-notch furnishings and shiny everything everywhere. Oh, and the salesperson was king for sure.

From the moment I parked my vehicle in the largest parking space I have ever seen in a garage (didn't want to scratch those BMWs, Mercedes, Porsches) to the moment I left that place in the evening, my stomach hurt. I never felt real there. I had a mask on the entire time.

I played the game, and I played it really well. I learned to act the perfect salesperson, yet, I am sure in retrospect that anyone who cared enough knew I was a fish out of water. I really didn't want to play this game, although the incredible financial and intrinsic rewards were enough to keep me there for a couple of years. Until I got laid off. Fired might be more appropriate. I was just beginning to question my instincts, and I wasn't just rolling over any more. I wasn't performing like the trained sales shark I had been for so many years. I was questioning my managers, questioning my heart. And, ultimately, I was let go.

Now, this was an absolute shocker for me. It was right before Thanksgiving, and I went home shell-shocked. I woke up in the middle of the night basically crying my head off at the front of the house so as not to disturb anyone else. I thought I had lost my identity, who I was, the big-shot salesperson.

What now? I had to be with my entire family in two days to celebrate Thanksgiving. (Little did I realize then that down the road a stretch, I would come to understand that this truly was a reason to give thanks!) So as I was bawling my self-pitying head off, I heard a cat crying. "Meow, meoooowwww" - a cat in distress. I forgot "me" for a minute and listened and sure enough, it sounded as if a cat was in the wall of the dining room. Long story short, my cat had come through the breezeway and fallen through the wall. I won't get into what it took to rescue that cat - which we did - but what a gift that was!

It got me out of myself, made me realize - if only briefly, but it was a glimmer! - what was important in my life, and the net of it was I am NOT a sales shark. I am not a hired gun who wants to take no prisoners.

I wanted to just be my real, authentic self and listen to my heart. Listen to what was real and not question that THIS was who I truly am. Be confident in myself, my true self.

So, if this book serves a purpose to help all the recovering sales sharks LISTEN to their true selves, then I am happy.

Transparency

I would never say all those years in sales not listening to my higher self, my instinct, my spirit and my soul were wasted because I believe that we are all here for a purpose, and I needed to learn those lessons to be able to share my experiences with transparency.

I was always naturally real, honest, transparent, but I allowed myself to "hear" that this was not the way to be. Being real was just not in vogue. I'm not sure it's any more in vogue now - although successful social media use is encouraging authenticity like never before. But I do know this: If we all embrace who we are and "be," just be who we are, well, the world would be better served.

Even when I was doing very well in sales as a sales shark and using my gift of communicating with all people, it was always about relationship-building for me. I have always excelled at that. 80% of my work during my sales career was starting and building relationships with business partners and resellers around our products and services. This spoke to me! Being social was what I was about. Listening, trusting and being transparent garnered many a client and friend.

It's the sales quotas that got in the way. The fear of lack. The "coming from scarcity" vs. coming from abundance that made me fear-based. Just recently, I met a salesperson who was just so sunny and open and real, and I wondered, Hmmmm, this person must not be a salesperson because they just come from abundance, as though their "quota" was made and they could just be the wonderful person they were.

I came to find out that the company they work for doesn't believe in quotas. They want their reps to represent the company without the pressure of a quota. The difference it makes in this "sales" rep's relationships is astounding! Everyone wants to be/talk/think and sit with this rep who comes from abundance vs. need.

I was so motivated, I now include a story on this company in every presentation and training class that I do. The company is the largest email marketing company on the planet.

So, all of this transparency has led me to this book on shedding the sales shark. This book's goal is to bring an awareness and change to sales professionals, encouraging them to be real, to be authentic and transparent and to listen to and trust in that inner voice, to listen to their heart.

I believe there are a lot of sales sharks out there just waiting to shed that shiny sales skin and be the true, authentic people they were born to be. So, let's say this chapter is dedicated to RECOVERING SALES SHARKS throughout the world!

Top Tips I Learned About Sales as a Sales Shark
You do not have to sell if you enroll people in what you believe in.
People want to grow relationships.
People want to be educated on what they don't know.
People want to trust in you.
No one wants to be sold anything.
Nobody likes to be made to feel less than.
Everyone wants to be heard/listened to.
Everyone wants to participate.
Everyone wants to feel special.
Everyone wants to feel loved.

Hmmmm, so many of the above statements are what social media is all about. It's about building successful relationships and engaging with people so they feel worthwhile and productive members of the community. So, whether it's blogs or Facebook or LinkedIn, it's all about the 10 tips above.

Business behaviors really haven't changed since the days of selling snow cones when I was a kid. People want to engage with you and relate to you… They want to feel part of the circle and in the know and smart and wanted and appreciated!!!! Nothing has changed. What has changed is the amazing means we have to make this happen on a global scale through social media! We have the ability to create dialogues vs. monologues!

I've just shared the top 10 tips I learned about sales during my time as a sales shark (or should I say, "a guppy in shark's clothing?"). Now I want to share the top seven universal lessons I've learned about **LIFE** in my journey from sales shark to recovering sales shark.

These lessons created the foundation that enables me to use social media successfully for my business and for my clients' businesses. Everything in life and social media first starts with you!

Lesson 1: Give to Grow

That's right, not "Win-Win." Not "Give to Get." How often we give with an expectation of getting something in return! My road to recovering sales shark has shown me that doesn't work. Give to Grow and you will get. The key is to know that there is a **"Law of Abundant Exchange."** This law means that when you give selflessly from goodwill with no selfish motive, you automatically get what you need most returned to you as a positive boomerang of your selfless generosity.

A corollary to Give to Grow is Give Back. Help those less fortunate than you. Whether you give back financially or with time and understanding, it's important to remember to take the time to help others. Why? Because it's one of the ways you get completely out of the way. In my experience, when you're giving, it's so pure, it's like you don't exist, time doesn't exist and it feels like something bigger, higher and more noble is going on.

Can you imagine letting go of worry about bills, and making sales quotas, and closing deals because it's the end of the quarter and doing things for the wrong reasons vs. the right ones? Can you imagine totally coming from a selfless agenda and just serving all we encounter?

Lesson 2: Get Out of the Way by Listening to Your Inner Voice, Not Your Mind Talk

The inner voice I am referring to here is the voice that comes from your spirit, your heart, your higher self that knows what is right and what is wrong. Call it your intuition or your gut. Sometimes it is just very hard to listen to, even though we know it's right because of how we feel when we notice this voice.

Many of us have TONS of mind talk going on. That's the continual dialogue of basically why you can't do something or shouldn't do something or how you're not smart enough or good-looking enough or rich enough and all that mess. This is not the voice you want to listen to. It is the voice that will hold you back and never let you prevail.

I think the biggest lesson for me in shedding my sales shark was realizing my ego was not in control. When I started my business, the first seven years were spent still using some of the operating beliefs and principals of corporate America. This was what I knew and trusted in. I hadn't quite grasped that the "higher source Shelley" was inside and available to guide me in all I do. Granted, I was well on the path to trusting in myself and my higher source; however, I was still motivated by financial gain. That is really the only measuring stick I had known for 25 years. There was no Give to Grow happening in my life. I was still coming from scarcity vs. abundance and the Give to Get mentality vs. Give to Grow.

I also was focused on helping other businesses thrive (not my business) for the first seven years of being an independent business owner instead of focusing on growing myself. I was "safe" working with companies and raising money through venture capitalists, angel investors and investment bankers. I could hide behind this façade of helping other businesses. I talked the talk and walked the walk, so I stayed in this most uncomfortable "comfort" zone since I didn't know how to let myself thrive.

I wasn't listening to what my purpose was yet. I used to say I didn't know what my passion and purpose were. I would ponder that and question, How does one find their passion and purpose? I think we all struggle with that. What I didn't do was LISTEN to the voice that guided me and pretty much was telling me what worked and didn't work. When I was helping people, training children, being my real self, those were the times Shelley got out of the way and my higher self came through – the self I am today, the self that is successful beyond my wildest dreams. (And, by the way, I finally got that there are other measures of success even more valuable than money.)

Back in the early days of working for myself, I helped many start-up and early-stage companies thrive in sales and marketing and raising money, but I still wasn't fulfilled. It just didn't feel right, it wasn't who I was.

Not until I decided to get back to my roots of training others by listening to my inner voice did the world explode in light and color and growth. In 2007, social media was just getting started, and I jumped in head first. I did a lot of research and created curricula around all the top social media tools. I also worked with many companies on their social media marketing strategies.

I knew I was home. The first class I ever did was a sell-out. We had over 75 people for a Social Media 101 course. I was really not prepared for what happened in that moment in time. When I was up in front of those people, all of them looking to me for guidance and training, I got out of the way.

The best way to explain this is that "Shelley personality" just stepped aside and let the knowledge and spirit and guidance to give come from outside of the personality self or what the psychologists call the "ego self." When I tried to "think," I stumbled and became fearful and was not effective. When I let go and trusted in all of the preparation I had done, and let the delivery come from a higher source than "little bunny Shelley," well, that is when the magic happened. That is when I realized that the path before me was the right one for me.

What happened on that stage giving my first social media training class was that I could be totally real, just be who I was, without judgment or concern about being right. I could just be, and WOW, what an amazing corner I turned! Seven years and hundreds of classes and speaking engagements later, I know my purpose is to be real, to be who I am without interference from my ego. I know this because people resonate with this real person, this real soul, and that's what they want. Many of us have wasted many years striving to be someone and something we are not.

So, how I arrived at this point was through the wonderful medium of social media that rewards those who are authentic and come from a place of truly caring, adding value, striving to be helpful and giving to grow. I have thrived personally and professionally from listening and allowing the real me to take a stand against the old "money-is-the-only-measure-of-success" mindset.

If you have that small voice trying to speak to you about what your passion is, where you thrive without thought, then listen. Sometime it takes many life lessons to reach this point. However, along the way, be in tune to what your physical body is trying to tell you. Feel what it feels like when you are coming from your sales shark vs. your higher self. It takes some practice, but you will know.

And in retrospect, you will realize that every river you crossed and mountain you climbed and path you traveled were all there to bring you to this point. Have trust. And give the same back.

Lesson 3: Realize the Insignificance of Your Significance

We think we are so "all that." We aren't really. When we realize the reason we matter is not individually, but as a part in the whole of humanity, then being a sales shark is a hindrance, not a help.

We, as humans, rarely look up. I mean really take the time to stop what we are doing and observe what is above us. We look down, we look left, we look right, we look straight ahead. But we seldom look up!

So what? Who cares? What does this have to do with social media and sales sharks? Well, for me, it became life-changing when I started looking up. And, additionally, it has given me an amazing perspective on the insignificance of my significance. OR, if you prefer, the significance of my insignificance.

I believe that when we get out of our own way and start to see our self as part of the larger picture in the world and universe, we can begin to become humble, which in itself is a huge lesson in humanity, giving to grow and excelling in social media! You see, this all ties together. If this book helps you shed that sales shark and gives you the ability to see yourself in both the digital and physical world as a positive force impacting all you encounter, then the book has done its job.

Okay, back to looking up. Once you look up, you can't go back.

Here's the story of how I started looking up:

One spring day, I was lying on the deck, feeling the warmth of the sun on my body and melting into the heat-filled wood. I looked up and what I thought was a common squirrel's nest was actually a bee hive, right there in a live oak tree about 60 feet above my head. Now I have had bee hives in my screech owl box, and I've had to have beekeepers come out and collect hives twice in my lifetime, but I had never had a bee hive fully exposed in a tree!

It was an incredible gift to be able to watch the bees form this hive and protect the honeycomb. When I got a glimpse of the honeycomb, which is not often as the bees usually entirely cover the comb, it was a treat to see. I got to watch that hive for several years until the bees vacated. *The one big lesson I learned from that hive is that we are nothing without each other.* We would not exist without our fellow humanity. And social media has been a boon to connecting people to join "bee hives" and contribute to the good of the world.

Every morning, I make a practice of looking up when I am outside. Whether it's taking the dog for a walk or getting the paper (yes, I still read a newspaper that's actually printed out on paper!), I make a point of looking skyward and without fail, I am given a piece of nature, a reminder of my humanity and my place in the world. I have seen giant herons crossing the sky, amazing sunrises that are humbling and egrets, owls, night herons, hawks, whistling ducks, a kettle of hawks and many other fine feathered friends flying overhead, which I would not have seen had I not chosen to look skyward.

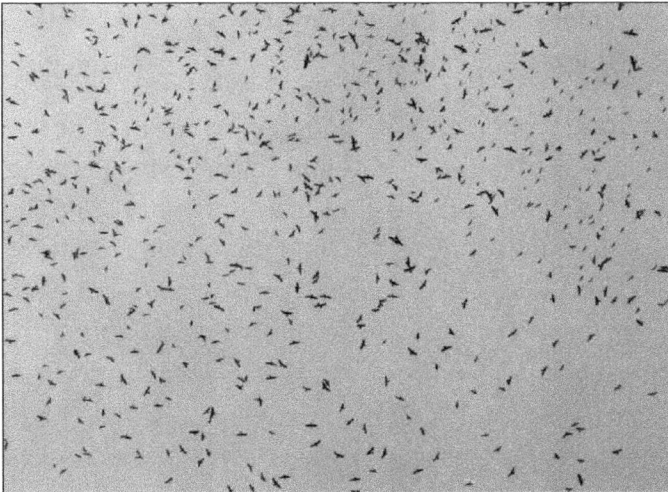

Photo Source: The Naturalist's Corner. Amazing kettle of hawks. Can you imagine seeing this many hawks at one time in one place?

Looking up is always a reminder of the insignificance of my significance.

It's just amazing the gifts that are available when we just look up. No matter your age, income, sex, gender......just look up. It will change your world!!

In the daily routine of things, it can be easy to think that what we do or how we do it doesn't make much difference. We are here for such a short amount of time in the scheme of things, so we can sometimes fall into the trap of "why bother?" From my experience, it's very easy to fall into this pattern of "so what, who cares" and live our lives numb to the good we can do as part of the whole.

This doesn't come easily. We are bombarded with media and commercialization and making it all about ME. Getting more stuff in hopes of feeling better about ourselves. Well, it's not about acquiring things. It's about contribution, and we all can contribute something to the betterment of our planet. And, it doesn't have to cost a thing.

A smile can rock someone's world. During a TEDx event in Sugarland, Texas this year, one of the speakers issued a challenge: Do something new for 30 days - whether it's giving up sugar, turning off the TV, reading a book, whatever you pick - do it for 30 days. I have been contemplating this and have decided to pick a date and smile at everyone I encounter for just 30 days. It seems like such a huge commitment to do this, yet it's just 30 days. I will report back on how this goes. I already know it will be life-changing. Doing anything for 30 days usually is. So, give it a go. Give up worrying about closing the deals or thinking you have competition. OR, smile at everyone for 30 days!

So how can we apply the lesson of realizing the insignificance of your significance to shedding the sales shark skin? One of the most effective practices is not worrying about closing the deal. When I go on sales calls now, I am very prepared. However, the biggest preparation I do is not having any investment in the outcome. That's right, I don't care if I close the deal or not. I know that the energy I put out will be positive, confident and attractive, and if the person I am sitting across from "gets" that from me, we will do business. We will do business if he or she resonates with my energy. Simple as that.

I finally gave up my sales shark and guess what, it works! I am working with people I want to work with and who want to work with me. Sometimes I have to walk away because it just doesn't feel right. Listening and trusting in this barometer was not easy at first. Why would I walk away form a *Fortune* 100 company and many $s? Because it just didn't resonate with me, it didn't feel right. This was hard to do, but oh so right.

So, walking away, saying no, listening to my inner voice that is peaceful and loving and comforting and remembering to look up at the sky, the clouds, the stars, the moon, the galaxy and beyond, I am reminded of the significance of my insignificance, and I am humbled. And, my sales shark is becoming a distant memory.

Lesson 4: Being "Flawsome" is Awesome

One of the coolest things about using social media for marketing your business is that you do not have to be perfect. As a matter of fact, being perfect is not trusted. Studies have been done that show that a restaurant or hotel with nothing but perfect scores is not trusted. People don't believe anyone or anything can be perfect. And the great news is that none of us are perfect. We are all imperfect in our perfection! Or, as I like to say,"Flawsome" is the new Awesome!

In the world of social media, people want to know you. They want to see that you are just like them and that they can relate to you. I learned this through doing my video newsletter, "The Social Media Tipster," which I send out weekly to my subscribers.

My only goal with my newsletter was to help people get better at using social media by providing short videos on some new development in social media technology. It wasn't to look good on camera or drive more business to me or aspire to sell my training videos. No, the truly bottom line was to give back. That's it: No hidden agenda.

Learning to use YouTube and Screen-cast-o-matic to create the videos was easy enough. However, I had no intention of sharing my face on these instructional videos. I didn't want to put my "flawsome" face out there and worry about "looking good" and all that goes with being on camera. I also didn't care about editing the video so it was perfect. If I made a mistake, so be it. I just wanted to share a teachable topic.

After making these training videos available, many people who received them were appreciative, but they told me over and over again that they wanted me to include my face on the training video. Ultimately, I relented since it wasn't about me; it was about providing value and being responsive to my community. I started including my mug on the training video. Whether I had on a baseball cap due to a bad hair day, or I included Buddy the dog (Springboard's mascot) on the video, I received so much positive praise for these "flawsome" instructional videos! What I realized once again was that people want you. They want the real you, the relatable you, the person with the awesome flawsome! Lesson learned.

So, don't worry about being perfect out there in the virtual land of social media. Just be yourself, without the agenda. Whether you are writing a blog, sending an email, posting a status update, tweeting, sharing content or doing a video, just be you. Be real! This is what people want for themselves. Reflect that for them and show them it's okay to just be.

Lesson 5: Believe in YOU

You must love yourself and like yourself before anyone else can. I recommend that you put aside 30 minutes a day, each and every day, to think about what you LIKE about yourself and write it down. Focus this time just on you, no interruptions, stay focused. Look in a mirror and tell yourself everything you like/love about you. Don't be negative!

A large part of being able to believe in you is really knowing who you are – and who you are not! In order to believe in ourselves and grow we need to be able and willing to assess our progress. This means knowing our strengths and knowing the areas where we have room for improvement.

I'd like to share two exercises in this section to help you know and believe in yourself.

Exercise #1: The InnerView Process

As we travel through life we seem to forget that there are at least four interpretations of who we are in the world. Johari's Window, which I presented when getting my Masters in Psychology, breaks it down for us in the chart below. (I actually keep a photograph of me doing that presentation on my desk, hair down to my waist and lilac bell bottoms and all!)

THE JOHARI WINDOW	Known to Self	Not Known to Self
Known to Others	1. Open	2. Blind
Not Known to Others	3. Hidden	4. Unknown

Source: Luft, Joseph (1969). *Of Human Interaction*. Palo Alto, CA: National Press, 177 pages.

I have found the Johari Window – named after the first names of its inventors, Joseph Luft and Harry Ingham – to be one of the most useful models for describing the process of human interaction. The four-paned "window" above divides personal awareness into four different types: open, hidden, blind and unknown. The division between the four panes is constantly expanding and contracting, depending on our interactions.

In this model, each person is represented by their own window. Let's describe mine:

1. The "open" quadrant represents things that I know about myself and that you know about me. For example, I know my name, my hair is brown, I am a short person. You know these things, too, when you've met me. After reading my bio, you know I am a teacher, salesperson, entrepreneur, etc. The information that this window represents is not only factual knowledge, but also my feelings, motives, beliefs, wants, desires and any information that describes who I am. When I first meet someone, the size of this window is not very large since little information has been shared. As I get to know you better, the parameters of the box shift so that more information is revealed over time.

2. The "blind" quadrant represents things that you know about me, but that I am unaware of. So, for example, we could be having lunch together and I unknowingly have gotten some food on my shirt. This information is in my blind quadrant because you can see it, but I cannot. If you tell me that I have something on my face (I would be most appreciative!), then the information shared moves it out of the blind spot and into the open quadrant's area. Now, I have blind spots with respect to many other much more complex things. For example, perhaps in our ongoing conversation, you may notice that I am not staying focused on you. Maybe it's my eye contact or listening ability, but something seems to be lacking. You may not say anything since you may not want to embarrass me, or you may draw your own inferences that perhaps I am disinterested in what you are saying or that I am being insincere. I will not be aware of this blind spot unless I am fortunate enough to discover it through a friend or colleague sharing or possibly through the InnerView© Exercise, which I will describe shortly. Uncovering these blind spots is very important in contributing to our continued growth process. To uncover things that others see that we are unaware of is a true gift IF we are open to learning and growing and developing what I call our "vision muscles."

3. The "hidden" quadrant represents things that I know about myself that you do not know. For example, I may not have told you that I love nature and one of my favorite ways to relax and connect with spirit is observing wildlife in my backyard. This information is in my "hidden" quadrant until I share it with you. There are vast amounts of information that have yet to be revealed to you. The InnerView Process is about others sharing what they observe, including our blind spots. However, it often leads to open discussions around areas that have been hidden from view by you.

4. The "unknown" quadrant represents things that are not known by me or by you. Being placed in new situations often reveals new information not previously known to self or others. For example, I might attend a workshop where something is triggered in me that gives me a new bit of knowledge that I didn't have and that others also don't have.

Through the Johari Window, you can see there are vast amounts of information you may or may not know about yourself and that others may or may not know about you.

The InnerView Process is an exercise to help you uncover blind spots so that you can determine how you are being perceived in your world. Benefits of this process include:

- Enabling you to acknowledge how you are being perceived by others
- Letting you release certain types of behavior that are not benefiting your growth
- Letting you apologize or be forgiven for certain behaviors or enabling you to forgive others for their behaviors
- Clearing up anything that may need to be cleaned up

Here are the steps in the InnerView Process:

First, establish a level of trust and freedom of speech with the person you are interviewing. If you are asking someone in your business community, then say, "What would the business community say if I weren't here?" or "What would our co-workers say if I weren't here?" or "What would our family say if I weren't here?" You want your interviewees to feel totally free to share their valuable feedback with you. You might want to tell them you are doing this for a project you are working on at work or school. Anything that will put them at ease during this process. Now, I encourage you to ask the questions and let them answer without any feedback from you, other than encouragement. Do NOT try and guide them. Just let it flow. After they answer, you might say, "Anything else you'd like to share?" or "Anything else you would like to add?"

Here are the specific questions to ask:

1. What are my strengths?
2. What are areas where I could improve? (With my Give to Grow philosophy, I prefer this language to the word "weakness.")
3. What does everybody recognize about me?
4. What can people count on me for?
5. What can people rarely or never count on me for?

InnerViews are a great way to view your progress over time. You may decide to interview one person a week for 6 weeks or interview 6 people in one week and then come back in 6 months and interview them again. It is very interesting to see how others perceive you before you started exercising your vision muscles through this process and after. At the very least, you are likely to uncover many blind spots as well as strengths and areas for improvement that are perceived by others.

Exercise #2: Moment of Choice

We have two ways of being in our lives: We can be proactive or reactive. Which do you think is more difficult? Right, being proactive. Many of us go through our lives in a constant state of survival vs. willingness and although some would argue it is a great character builder, I am here to tell you that coming from survival is NOT a pleasant, enlightened, loving place to be.

In the Moment of Choice Exercise, we need to be able to recognize when we are losing power, freedom and self-expression so that we can choose another way of being. This is what is called "Moment of Choice." The steps to working through the Moment of Choice Exercise are as follows:

First, notice when you are feeling like someone has stuck a pin in you, kind of like a slow leak in a balloon.

All choices in our lives are decided in how we respond in the moment. The better you are at noticing or being present to your negative self-talk, attitudes, feelings and behaviors, the better you will become at exercising your control over them and developing your vision muscles. So, first we must **NOTICE** what is occurring and how it is affecting us. Honor yourself for noticing, take a deep breath and then choose again how you will react to it. Become present to the present!!!

The following Moment of Choice diagram shows the Moment of Choice being the present. When we fall below the line we fall into judgments, assessments, need, lack, expectation, disappointment, fear, hopelessness, resentfulness, etc. These all come from our Past and make up our Reactive Mind. When we are in our Reactive Mind, I equate this to being off-kilter, off-center, imbalanced.

On the other hand, our Proactive Mind is all about being loving, forgiving, fulfilled, perfect in our imperfection just as we are, non-judgmental, appreciative and grateful.

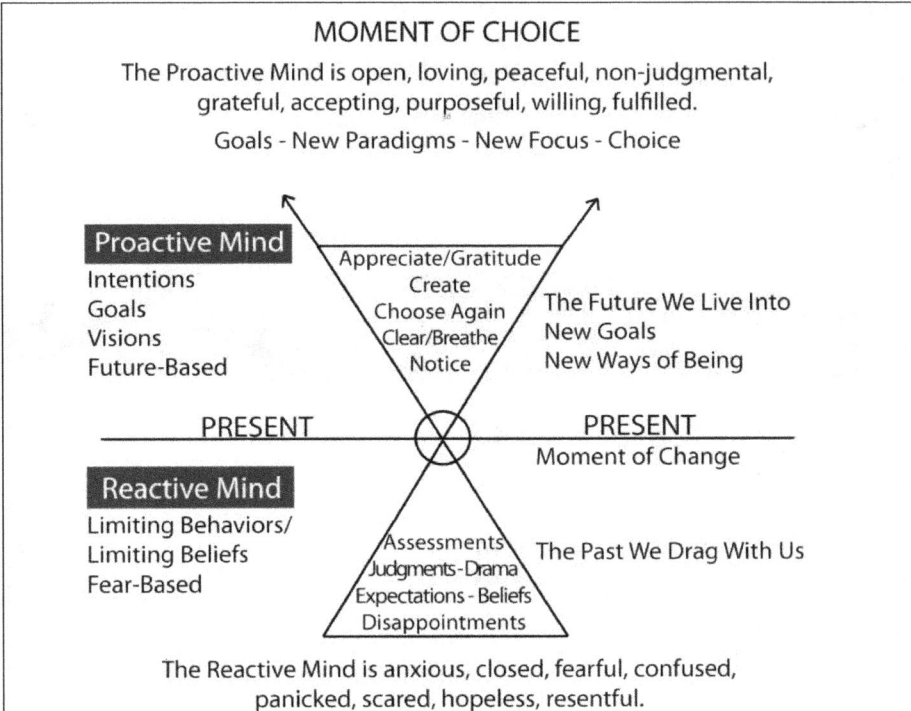

MOMENT OF CHOICE

The Proactive Mind is open, loving, peaceful, non-judgmental, grateful, accepting, purposeful, willing, fulfilled.

Goals - New Paradigms - New Focus - Choice

Proactive Mind

Intentions
Goals
Visions
Future-Based

Appreciate/Gratitude
Create
Choose Again
Clear/Breathe
Notice

The Future We Live Into
New Goals
New Ways of Being

PRESENT

PRESENT

Moment of Change

Reactive Mind

Limiting Behaviors/
Limiting Beliefs
Fear-Based

Assessments
Judgments-Drama
Expectations - Beliefs
Disappointments

The Past We Drag With Us

The Reactive Mind is anxious, closed, fearful, confused, panicked, scared, hopeless, resentful.

Now the challenge is to be aware of when we slip below the line, into the past, and we are feeling "less than." Once we slip below the line, we must choose again to come above the line and establish a new way of acting towards whatever pushed us below the line. As you can see with the circle in the middle, it is a constant struggle to stay above the line and focused on choosing again. However, it not only can be done, it must be done to get past the past that holds us back and keeps us from living a life of love and gratitude.

In this exercise, which is practiced every moment of every day, be aware of when you are going below the line, feeling small, unimportant, less than, and then be conscious and choose again how you want to react to the stimuli that pushed you down. We always have the power to control our mind about the situation and our response to it.

A Give to Grow Footnote: Many thanks to Kairos Foundation and Landmark Education for influencing the creation of the exercises in this section.

Lesson 6: Live Fearlessly & Have Trust

Letting go of plans, strategies and agendas is tough to do. As a sales shark, can you even picture that possibility?

In my own business as a social media navigator, I don't care about my business being number one in SEO (search engine optimization)...But the other side of my brain – the left-over prehistoric sales shark part – says, "Yes, I do. That's what a successful company does."

Then I have to remind myself to live from a place of trust and fearlessness, doing what I do because I feel it's right, doing what I do because I am called to, not because I have to be number one or promote any agenda. There is no agenda, just Give to Grow, follow my heart and share my knowledge so that people who want to have an agenda and strategy can use what I train them on to go forth and prosper.

For me it's an amazing place to come from, not judging myself or measuring myself or feeling I have to be number one or having to sell more. This isn't to say that having agendas and plans is a bad thing. I think we need to know where we are going if we want to get there.

One of my gurus on Facebook stated that everything she does on Facebook has a marketing plan behind it. While I respect that - because I have been there, and I help my clients with how to think about marketing more effectively via social media - I actually don't want to have a marketing plan at this stage in my life and business. That is a choice I make. I don't want to worry about who is going to buy something or who won't buy something. The more I let go of having a marketing plan and having to sell something and having a strategy to sell, the more I attract to me! This has been a huge test for me of whether I've actually shed my sales shark.

Wouldn't it be wonderful if we didn't have to be so focused on having a strategy? Wouldn't it be great if we had more trust to know everything will happen as it should because of the Law of Abundant Exchange that supports Giving to Grow, because we have a Higher Self whose inner voice can guide us wisely, because we are part of a bigger growth blueprint for all of humanity that helps support us when we come from a place of doing well by doing good? Wouldn't it be great to just be prepared and know your stuff and then just give? Knowing and trusting that people contact us because they want what we have instead of cold-calling and spouting data and using manipulative, insincere ways to "get past no."

Can you imagine a world where we lived on trust and didn't worry about bills and making sales quotas, and closing deals because it's the end of the quarter? Can you imagine totally coming from a selfless agenda and just serving all we encounter? If you can even contemplate this, then you have the potential to shed your sales shark skin – or help others you manage or care about shed theirs.

Simply put, I've learned that we should have trust in how things will unfold. No fear of what could be, would be, should be. Fear is never good, whether you're face-to-face or online. People can smell it, and it is not something that anyone is attracted to.

My journey to today as a recovering sales shark started by creating a workshop called "Living Life Fearlessly" after some serious coursework with Landmark Education and Kairos Foundation. I declared myself a "public speaker" and from that moment forward, I was a speaker. I presented this work at many public speaking venues including Women In Energy, E-Women Network, Landman Association, Dress for Success. I also offered it as an all-day workshop.

Let me tell you, this was not easy. Who was I to think I could get in front of successful people and talk about living fearlessly?... Because my knees were definitely shaking and my "mind talk" was screaming at me that I couldn't do this. My mind talk was telling me that there will always be someone in my way. It could be myself, (which was the case for me) or a partner or friend or anyone that wants to hold you back from your passion and purpose.

The way I got my mind talk to quiet itself was just to realize that you are what you create yourself to be. You declare it, then you live it... Perception is reality. Perceive what you want to achieve. Don't hang around people who bring self-doubt into your head. Those are not the people you want around you to achieve your life's purpose.

After getting over my own fears, I knew I had found my "voice" and my passion when I was delivering the "Living Life Fearlessly" workshop. When standing in front of a room, little bunny Shelley miraculously got out of the way, and an energy flowed through me. An energy that I had nothing to do with.

Now, let me say I was always more than prepared for the work I was going to deliver. The more prepared I was, the more little bunny Shelley could stand aside and this newly discovered piece of me - call it trust, hope, my higher self or whatever else you want to call it - could talk through me. When I "thought" about what I was doing up there or what I looked like or anything other than being the messenger, that's when I didn't do well.

Springboard Tip

Know your work and then get out of the way and be led by a higher power. Call it "heart, source, faith, energy vibe, whatever"... It works!

Our Deepest Fear

Our deepest fear is not that we are inadequate.
Our deepest fear is that we are powerful beyond measure.
It is our light, not our darkness, that most frightens us.
We ask ourselves, who am I to be brilliant, gorgeous, talented and fabulous?
Actually, who are you not to be?
You are a child of the universe.
Your playing small does not serve the world.
There is nothing enlightened about shrinking so other people won't feel insecure around you.
We were born to make manifest the glory of the spirit that is within us.
It is not just in some of us; it is in everyone.
As we let our light shine,
we unconsciously give other people permission to do the same.
As we are liberated from our own fear,
our presence automatically liberates others.
- Marianne Williamson

springboard

I believe that if you are willing to focus on your dreams instead of your fears, anything is possible. Exercising your vision muscle is one of the best ways I know to start making your possibilities into realities.

Exercising Your Vision Muscle

How many of you go to the gym, run or do some form of exercise to keep "in shape" physically?

We spend hours at the gym, running and swimming to exercise our muscles and stay in shape. We do repetitions. Three sets of this for my triceps (wouldn't want that flab on the back side of my arm to show). This outward physical display, while very important to our physical and mental health and nice to look at, doesn't address what is happening to our vision muscles. We spend so much time on "How we look, looking good," that it never occurs to many of us to exercise our minds.

Now, some of you might be saying, I meditate and pray and go to church and read and this is all great stuff for our minds. What I refer to today, this vision muscle, can incorporate all of the things like reading, praying, etc., but it goes one step further. The vision muscle is the muscle that, without development, becomes weak and useless and holds us back from really doing some serious pumping of possibility....

This is the muscle that, if left undeveloped, is responsible for a lot of the fear muscles that occur in our lives. This fear is usually unwarranted fear or anxiety or nervousness or whatever occurs that leaves us feeling depleted of energy, weak, without power, little freedom, or not expressing ourselves and CERTAINLY NOT living a personal or professional life we ENVISION...

Let me clear up what I mean by fear. I don't mean a true fear, a trigger of a fight-or-flight response like stumbling upon a hungry lion or almost being hit by a car. I mean a "made-up" fear that might have been created when we were young and latched onto with the result that it developed a "fear muscle" in our memory. For many people, this fear muscle reasserts itself in our memories when various triggering situations occur in our lives.

Now developing a fearless "yes" muscle or vision muscle is a matter of repetition, just like at the gym.

First, let's figure out when this FEAR muscle was first developed and when it is exercising its power over you. I ask you to consider that when you feel a loss of power and energy, etc., something occurred a long time ago that created this response.

Studies have shown that mind talk is 60-90% mostly negative!!! In addition, studies have shown that 96% of the input we receive as children from parents, teachers, coaches, etc. is negative. That means that 24 out of 25 comments we received as children were negative....No wonder for many of us our vision muscle is totally out of shape, and our fear muscle rules the day!!!

What we tell ourselves (we aren't smart enough, pretty enough, worthwhile enough, rich enough, thin enough) ...all this mind talk from this little voice... tends to run us off and leaves us totally uninspired and unempowered.

Interpretation vs. Truth: It's All Made Up!

Consider that much of who we are comes from the past and we are weighed down by what the past has dictated we are - baggage we carry with us, stories we make up, interpretations of stories. Attitudes and beliefs often affect actions and ultimately who we are and how we either move forward or stay stuck.

When I challenge myself to identify my various fear muscles - those disempowering, lack-of-energy, heavy hearted muscles - I first must try and figure out what created it in the first place (which takes some great coaching from a trusted friend or hired coach; see Lesson 7 later in this chapter). Then I have to develop or invent a new way of exercising my vision muscle so that my response is different. The vision muscle grows and develops around this new behavior I have created that I want to occur.

Let me add here that I've experienced that some people just don't want change. They just want to remain on autopilot and go through life status quo. Maybe they aren't in enough pain and suffering to change OR there is no excitement around change OR the fear of the change is greater than the excitement of the new possibility. For these people, they are getting some kind of payoff for holding onto whatever has them stopped.

But if you really want to change and shed your sales shark skin, if you are tired of your beliefs and attitudes holding you back from exercising your "vision muscle," then, the first step is:

1.   **NOTICE.** Recognize when you are losing your self-expression, energy, vitality. Acknowledge it AND choose again. (Breathing helps!) Decide what you DO want. Here's an example: I used to say NO to every opportunity to speak in front of groups because of this dread and fear I had from my little bunny Shelley

experience getting shocked on stage. These early experiences are all part of who we become; some give us winning ways of being, some hold us back. Anyway, I would no more stand in front of a group 10 years ago than I would volunteer to be shocked by that microphone again. Today, I'm in front of groups all the time. It took a lot of work on my vision muscle to get to where I am today, but it was worth it!

2. **DECLARE YOUR INTENTION.** The second step in developing your vision muscle is to invent...create....declare... what it is that you want for yourself, personally and/or professionally. Create this future without any restrictions on what is and isn't possible or what you can or can't do. Once this is created, then declare it both internally and externally. Then go live it step by step.

3. **IDENTIFY THE GAP.** This third step is when you will determine where you are today (the present) and see the gap and what needs to occur to bridge the gap to get to your declared vision of tomorrow. To identify what you need to bridge the gap, ask yourself, What would it take to go from a 1 to a 10? So you want to become a public speaker. What has to occur to make it real? You have to commit to measurable, actionable results. For instance, you might sign up to talk to various groups or join Toastmasters.

4. **BE ACCOUNTABLE**. This is the fourth step. You must be your word to yourself. This is by far the hardest step because it is so easy to NOT be accountable to ourselves. We are accountable to others without a second thought. However, for some reason, when we give our word to ourselves, we break it ALL the time. If we did the same to friends, clients, co-workers, most of us would consider ourselves out of integrity. Being our word is important not just to others, but also to ourselves. Now I challenge you to

decide how to keep your word to yourself. When you make a declaration to you, keep your word.

The MAAP Process for Accountability

A MAAP© (Measurable Actionable Accountability Plan) is a written program that brings you results when you work your plan. It tracks your progress through the course of your declaration and commitment. It includes actions you must take to produce the results you are committed to, accompanied by a coach to hold you accountable to your word.

Here are two important things to do as part of the MAAP process:

1. Create and write down the future intentions you want to occur. Envision them. See yourself in them. Write down: How do you look? How do you act? Who are you with? Where do you live, etc.?

2. Keep your intentions in existence by having a support structure. Surround yourself with people who will be positive around the results you are committed to. Tell someone about your results and future; then ask them to tell someone else, then someone else... Viral marketing at its best! It all starts with putting it out there and making it a reality. Make a poster or collage with pictures you cut out from magazines, or use Pinterest to make a vision board online. Put the poster where you will see it often. Then, show it to someone and tell them what it means to you. The more people who know about your intended results, the more likely you are to act consistent with that commitment you have for the future.

Questions & Exercises to Help You Prepare Your MAAP

- **What desires and dreams are still unlived in you?**

- **What actions do you need to take to unleash them?**

- **Acknowledge/appreciate accomplishments.** Acknowledge/appreciate your accomplishments over the last 12 months. What are you proud of? What did you do well? What did you complete? Examples: spoke at an inventors association, attended three self-improvement workshops, ran a marathon.

- **List disappointments.** What hurt you? What didn't go as planned or hoped for? Tell the truth. Examples: I didn't spend enough time with family; I didn't look up every day like I promised I would; I didn't listen to my inner voice.

- **Insights from accomplishments and disappointments.** Ask yourself, what lessons did I learn from the accomplishments? What do I see? What have I learned from the disappointments? Make these insights positive-memorable-short. Examples: I want to work with others; I want to be paid for my work; I want to make a difference in people's lives.

- **Pick your top 3 goals/intentions/results from Insights.** These will go into your MAAP.

- **What are your limiting paradigms or beliefs, or what has stopped you?** How do you limit yourself? Put down what pops into your head. List 3-10 limiting beliefs.

- **Create a NEW paradigm.** This must be positive, personal, present tense, powerful, simply stated and reflective of an exciting future for you.

All of these exercises give you room for thought to prepare a MAAP.

So, are you ready to create your MAAP now? If you feel ready, give it a try using the following form.

MAAP = Measurable Actionable Accountability Program

This program translates activities into measurable results to produce a direction and accountability for you and what you are committed to achieving in your life.

MAAP Plan SMART Goals: Specific, Measurable, Actionable, Realistic, Timed Goals for:

_____

Your Name and Date

I.   **Results I Will Produce:** Concrete, measurable, achievable, yet a stretch

List the results you want to occur for the next _____ months.

_____
_____
_____
_____
_____

II. **Purpose:** Why is this result (outcome) important to you?   For every result, list at least 1 purpose.

_____

_____

_____

_____

_____

_____

_____

III. **Visioning:** Actual visualization of this result occurring (write them down). What do you see in this result? Be specific.

_____

_____

_____

_____

_____

_____

_____

IV. **Agreements:** Specific action steps that will create your result with dates and timing. (Must have dates and times associated with each result.)

1. _____

2. _____

3. _____

4. _____

5. _____

6. _____

V. **Support Coach:** Select someone who will hold you accountable to your plan. Weekly contact is preferable. Schedule a recurring event to connect with your coach on your calendar, and ask them to do the same. Give them what they need to support you, and hold them accountable to their commitment to you!

Plan Your Work….Work Your Plan!

Support Partner Name: _____

Partner Phone: _____

Partner Email: _____

Springboard Tips
There is no right or wrong;   only thinking makes it so.
Would you rather be right or happy?
Perception = Reality.   What we perceive, we achieve.
One of the things you have the power to change is your   mind…Choose again how you respond.
There are two major fuels of being: fear (anxiety, worry) and love (peace, harmony).
It's not what you do,   but who you are being when you do it.   We are Human Beings, not Human Doings!
You are the author of your own Life Script.
Rearview mirror thinking stops us   from our dreams.
Nothing real can be threatened,   nothing unreal exists.

I'd like to close this lesson on living life fearlessly with one of my favorite stories:

> *An eaglet fell out of its nest. A farmer found it and put it in with the chickens, so this eaglet grew up eating with the chickens, living like a chicken, pecking around the barnyard like a chicken, doing what chickens do. It grew up with a chicken consciousness. It perceived of itself as a chicken, and no one told it any differently. It had wings but didn't fly because it thought it was a chicken. One day, the eaglet was out in the barnyard pecking around when a mature eagle flew overhead. The eagle looked down and called out to the eaglet. The eaglet, now a mature eagle itself, looked up and saw an eagle for the first time. And something happened in that moment to that eaglet still living inside; it was called toward something higher. And so this eaglet - now grown - heard that call and something awakened within it about its true nature. Identifying with that true nature, it became something different and it flew and became the eagle that it was intended to be.*

It's a wonderful story because if we look at our own lives, we all have the potential to fly off and become the eagle that is within us.

What is so powerful about the notion of becoming who you are is that who you are is who you ultimately say you are, who you declare yourself to be. You make a declaration, and you create the possibility of being (fill in the blank) _____.

Maybe you grew up learning certain so-called truths about yourself and what you could accomplish – and you believed what you heard as if THOSE LIMITS were the truth.

But if you listen for the call of your own true self and are not afraid to fly towards it, everything you can imagine and much more that you can't yet imagine is possible for you.

A Give to Grow Footnote: Many thanks to Kairos Foundation and Landmark Education for influencing the creation of the MAAP.

Lesson 7: Get an Accountability Coach

I think it's imperative that we all have an accountability coach, someone that holds us accountable to our word to ourselves until we have developed and exercised our vision muscle enough to keep our word to ourselves. As I mentioned before, surround yourself with people that support your vision muscle. This might mean shedding some of your current friends or even your current job, along with your sales shark skin.

The accountability coach will help assure you stay on track with your goals. It's one thing to promise and commit to ourselves that we are going to accomplish X. It's very easy to get sidetracked and not hold the promise we make to ourselves. I mean, who will know that we weren't our word except us? And by the way, this is huge. You are most important. You must be your word to yourself. Having an accountability coach is the first step in helping make this happen.

An example of working with an accountability coach is creating my first book, *GET REAL, People!* Without the recurring meetings with my energy arts advisors to keep me to my word on delivering this book in a predetermined timeframe, the book might never have been published.

Even more importantly, when I set out to write my book with the help of these accountability coaches, I thought it would just be about editing and keeping to deadlines. However, it was about so much more. Even though my energy arts advisors were serving as editors and design visionaries, what they really did was help me discover more about myself through the book writing process than I knew. They saw way beyond the technical aspects of the book and helped me realize that the book was as much about expressing my voice and helping others grow by sharing that voice as it was about writing well and staying on deadline. They helped connect me to my truth, and the feedback I've gotten about the first book is that one of the results for those who have read it has been that they connect to more of their own truth. Give to Grow in action!

Another example is working with an accountability coach to exercise or lose weight. These are pretty typical areas to have help from someone to hold us accountable. I have lost many a pound and worked out for over 25 years of my life, and it all started with an accountability coach.

Interviews with Corporate People
Who Left the Shark Tank

As you can tell by now, I love stories! As an educator, I discovered that some of the best learning happens through stories as compared to memorizing lots of facts. Social media is all about stories and the ability to share them, and that's one of the reasons Facebook has just surpassed one BILLION users.

I didn't feel that this book would have been complete with just my story and lessons. Since I love engaging with people, giving to grow and helping other people shine, I wanted real-life stories about others who have "left the shark tank" to be part of this book's experience.

I have often wondered over the years how other sales sharks survived after leaving the corporate waters. My definition of "sales shark" isn't limited to those in sales. As I mentioned in Chapter 1, I use the term "sales shark" really to describe anyone who is in a job that may define them, mostly in a corporate setting, where creative thinking and feeling with the heart are frowned upon.

In my sales shark days, I experienced so many instances of doing things I knew didn't feel right inside my body and knew in my heart would not serve the betterment of humanity, but the corporate greed that I had come to believe was the right way became what I thought was the right thing to do.

This couldn't have been further from the truth. Heck, if I had followed my higher self – truly listened – I might have been fired sooner and begun my amazing journey earlier. However, all things happen as they should. Only later, after having my own business and running it as I saw fit did I realize that so many corporations are run on fear and greed and lack, how could they be healthy?

Just as we humans must be nourished physically, emotionally, mentally and spiritually, so do businesses. They are very much like human beings and need the same healthy environment both inside and outside their corporate walls.

With that in mind, I interviewed former corporate employees to see how their sales shark was doing after leaving the tank. Each person I interviewed agreed that it was the best decision they had made and that the experience was liberating.

Interview: Jill Hickman

What prompted you to leave working for a corporation?

My dad. He said that if I was going to work crazy hours, I might as well be doing it for myself!

How has it changed your life?

I would hope that I have changed the lives of others, and in that process, that has made me a better person.

Do you have any regrets?

No. I believe that everything happens for a reason, whether or not I like it or understand it. I do take each failure – in relationships, jobs, and projects (and there have been many!) – as lessons learned. I always look for the silver lining – the gift in the experience. I believe that each experience is like a puzzle piece. I may not like the piece; I may not see its beauty; I may not see how it fits into my puzzle at all. In fact, I may not even WANT that piece. But I know one day, that puzzle piece that sits there unconnected, seemingly unrelated, will fall into place with all the other pieces, and it will make perfect sense – in its own time. I love that I recognize that about my life now. I see every interaction, every project, every experience as another puzzle piece. It's an exciting picture that is developing.

What's the best part of having left?

I get to make all the decisions. That's also the worst part. I have no one but myself to hold accountable for a poor decision!

What surprised you most?

How much I love it. How doing what you love is more like play than work. People think I work all the time, but I really PLAY. I LOVE what I do.

Do you think there is a type of person that is more suited to make this transition and if so, what do you think made them suited to follow their passion?

Anyone can have a hobby or a passion and can START their own business from that point. What makes someone successful at their business for the long term is their ability to take risk, to market themselves and their business, and to sell themselves and their products/services. No one buys a product/service. They buy from people (or businesses) that they like and trust. If you can't get people to know you, like you and trust you, oops – work for someone else. But then again, I'm not sure that'll work for you in the long run.

Your Former Title: Director of Training and Development

Where were you formerly employed? Business Services at a business-to-business company

Please provide your current company or position: President, Jill Hickman Companies

Interview: John Graham

What prompted you to leave working for a corporation?

I got tired of the BS with owners and wanted to get back into consulting where I had more time flexibility and where I can truly contribute my knowledge and skills. I got tired of dealing with personality flaws.

How has it changed your life?

You live a bit differently without the guaranteed paycheck every two weeks.

But you also live freer without the stress of knowing you have to be there every Monday through Friday or else have a good reason why you are not there.

Do you have any regrets?

Not really. I do miss the secured retirement that can be offered by some larger companies. But even that is fading for most companies.

What's the best part of having left?

More flexibility to live life the way I want to live it.

I once read, "What price are you willing to pay for the money you make?"

What surprised you most?

How much consulting work there really is once you get wired in.

Do you think there is a type of person that is more suited to make this transition and if so, what do you think made them suited to follow their passion?

Yes, only those who have a strong belief in themselves. Also, you cannot be someone with lots of debt, living paycheck to paycheck. You need some resources to fall back on until you get established. It doesn't happen overnight, but you have to work on it every night.

Your Former Title: Chief Operating Office

Where were you formerly employed? Oil and Gas Manufacturing

Your current company or position: Frontline Resources, Inc,. www.houstoniso9000.com

Interview: Marcie Hysinger

What prompted you to leave working for a corporation?

I was in a sales and business development role when I was inspired to strike out on my own. Drawing on 15+ years' experience in sales, I began working with small and mid-sized companies to create and implement customized sales strategies.

How has it changed your life?

Personally, working for myself has enabled me to have a more flexible schedule and better accommodate "family time."

Professionally, it has allowed me to step out of my comfort zone and gain experience in a wide range of industries.

Do you have any regrets?

In retrospect, I wish I had outsourced more of the functions that were outside my core competencies, such as accounting and appointment setting.

What's the best part of having left?

Personally, the best part of having left is the flexibility. Professionally, the best part of having left is the ability to work nimbly, making decisions quickly and efficiently.

What surprised you most?

Sorry, nothing comes to mind here.

Do you think there is a type of person that is more suited to make this transition and if so, what do you think made them suited to follow their passion?

To be successful in this transition, a person must be realistic about their financial potential and the associated timelines. In addition, they must be very self-disciplined and have a thick skin to weather the ups and downs.

Where were you formerly employed? I worked for a global leader in press release distribution and regulatory disclosure.

Your current company or position: Principal Consultant at Pivot Advisors LLC

Interview: John O'Dell

I left the University of Houston after 8.5 years.

What prompted you to leave working for a corporation?

I conceived ideas for aligning business development strategies with ones' interests/passions and a Sales Relationship Management Software that would significantly impact sales professionals' productivity and enjoyment of the business development process.

How has it changed your life?

I have returned to an excitement level that I experienced when being a key executive of creating paradigm-changing Internet enabled software like the first PC postage, the first online ticket sales/print/use system and the first non-disruptive liquids tagging system.

Do you have any regrets?

No, I am still able to engage with University of Houston and continue to make a difference there.

What's the best part of having left?

Pursuing a stronger passion.

What surprised you most?

No surprises so far…

Do you think there is a type of person that is more suited to make this transition and if so, what do you think made them suited to follow their passion?

I am an entrepreneur who disciplined himself to work within a major organization environment. I know what entrepreneurship is all about and therefore am simply transitioning back. Corporate professionals who leave major companies (and all their resources) often have no idea what it is like to have to handle so many functions, even running the copy machine, at a small staff company.

Your Former Title: Director, Alumni Development

Where were you formerly employed? C.T. Bauer College of Business, University of Houston

Your current company or position: Chairman & CEO, Cink

Interview: Lexy

What prompted you to leave working for a corporation?

I thought I had the dream job at a large city newspaper as the Director of Talent Management. Six months later after laying off the whole Seattle PI, I received my layoff papers. From there, I did contract work and looked for an opportunity to earn equity in an organization that needed my talent. It took me two engagements like that, which did not turn into what they were sold as, to know I needed to create my own way.

How has it changed your life?

Everything that seemed impossible is now not only possible, but happening in ways I could not even have seen as a possible.

Do you have any regrets?

I have found regrets to be a waste of time….It all happens just as it should…You just have to keep moving.

What's the best part of having left?

I am able to leverage my strengths every day and help others do the same… It is why I am on the planet.

What surprised you most?

That I am able to process through the fears more rapidly than I thought I would be able to.

Do you think there is a type of person that is more suited to make this transition and if so, what do you think made them suited to follow their passion?

Persistence and resilience are needed. There will be more you do not know than you do, and you will need to be able to ask for help and create a network that enables you to do what you do well. Outsource the rest.

Your Former Title: Director of Talent Management

Where were you formerly employed? Media company

Please provide your current company or position: Managing Partner of Fokal Fusion, www.fokalfushion.com

Interview: Rob Longenecker

What prompted you to leave working for a corporation?

During my corporate sales career I'd had several opportunities to act as if I were an entrepreneur within the business. That's when I felt the most energy and passion. I started my current business part time on the Internet while still employed, so when downsizing caught me, I was ready.

How has it changed your life?

I have more passion for my work and love creating possibilities that energize me.

Do you have any regrets?

Absolutely none.

What's the best part of having left?

Despite the challenges, I like driving my own destiny and have developed the confidence that I could crash and begin again. I like standing or falling on my own, but still finding ways to cooperate with other business owners for mutual benefit.

What surprised you most?

It's really true that when you truly commit to a goal, the universe delivers the resources you need. Stuff just shows up. People show up, too.

Do you think there is a type of person that is more suited to make this transition and if so, what do you think made them suited to follow their passion?

It takes a person who will venture forth, stepping out on the skinny branches without knowing exactly how the path to success is laid out. Many people want their own business, but won't act until they have all their circumstances lined up just right. Then, "just right" never shows up. Small business isn't built on circumstances. Rather it's built on intention, persistence and a dash of intensity.

Your Former Title: Regional Manager

Where were you formerly employed? Loctite Corporation, Briggs-Weaver Industrial Supply, International Training Corp. and Osborn Manufacturing

Your current company or position: Owner of Tucker Leather, an Internet sales based company with customers all over the US (and some in Europe and the Far East)

Interview: Debra Tummins

After 32 years in the technology industry and more years than that of employment, I made the decision to leave Corporate America in 2010. Making the decision to leave did not come easily. But I realized at some point that I was putting so much of myself and personal growth and fulfillment on hold.

What prompted you to leave working for a corporation?

My father had clearly instilled in my mind at a very early age the concept that the value of one's life is measured by how hard one works. My personal and career goals were always those that involved working hard, working smart, making money, being recognized, winning and ultimately achieving professional and financial success. I had had an incredible and rewarding ride and am quite fortunate because of the opportunities I have had over the years. My decision to leave was most likely similar to many. I had reached a point where I believed that I was missing out on so many things in life because of my work. For so long my quality of life was measured by material gain and professional success, and I finally reached a point where I did not want my work or some executive title to define me.

How has it changed your life?

Stepping aside from Corporate Life when I did was by far the best decision I had made in years. I believe we all want to be inspired and challenged every day. This was not something I was feeling or experiencing. I have thoroughly enjoyed exploring new opportunities that I had put on hold for so long. I have been on a major campaign to invest in myself, better health, happiness, and overall fulfillment. The title and the job were my shield. Once that came off, it has been sometimes frightening but always freeing and rewarding.

Do you have any regrets?

No. Well, I might occasionally miss the nice pay checks. And I really enjoyed working with new and emerging leaders, mentoring and coaching them, but I know I will continue to be able to do this, just with a different structure.

What's the best part of having left?

Freedom, flexibility, better health and happiness, and, for sure, less stress. I still do not have enough time to do all of the things I want to do.

What surprised you most?

We are often defined by our professional roles. I certainly defined myself that way. When that title and status are taken away and we are standing there "bare naked" it can be a big adjustment. I had always believed that I had my ego in check, but going through this has made me realize that I still have some work to do. Without the job and title, every day is a re-invention. It can be scary and depressing for some. I choose to make it inspiring, exhilarating and, most of all, fun. What is also surprising is that it will sometimes change relationships with others. Sometimes others view you as that person with the big job. When you don't have it anymore, sometimes the relationship changes as well. It has been funny and strange at times, but I am okay with it.

Do you think there is a type of person that is more suited to make this transition and if so, what do you think made them suited to follow their passion?

Sometimes I wish I had made the transition earlier but I know we are all right where we should be at any moment, so I have no regrets about any of my past experiences.

Some people could easily make the transition if 1) they already know what they will be doing in their next act 2) if they have never allowed their job or title to define them and they see unlimited opportunities and options ahead of them no matter where they go or 3) someone who is on a self-fulfillment journey and they know when they have gained as much as they can gain from Corporate America. I would recommend to anyone that they have an idea as to what they want to do after corporate life. Corporate life was a wonderful ride for me for a very long time. Many people work their entire lives and are totally fulfilled, and I think that is great. For me, it wasn't enough. Inspiration and exploration were knocking at my door.

Your Former Title: Senior Vice President, North American Sales, CA Technologies

Where were you formerly employed? Technology companies from AT&T, IBM, BMC Software to CA Technologies

Your current company or position: Self-employed to work on my third act

Interview: Richard Janik

After a few short years working for a couple of marketing firms, I quickly realized that I had all of the skills necessary to start my own business and not have to depend on a company to carve out my path and future in this world. So while still employed, I worked tirelessly in my off time (nights and weekends) to acquire a handful of clients that I was servicing on a freelance basis. After only one year of sacrificing most of my free time, I was able to save enough money and establish residual income to quit my job (which was horribly stressful and no fun to work at) and work for myself!

What prompted you to leave working for a corporation?

Two things: The first thing that motivated me to leave was the fact that I was overworked and underpaid. At first I did not mind since I loved working as a multimedia specialist, but that brings me to the second reason I quit - being very unappreciated combined with typical office politics, lies and deception.

How has it changed your life?

My life has changed close to 1 million % for the better. Every year since I started my online marketing firm, SSD, Inc. (www.fairmarketing.com), I've managed to double the company's revenue and profit so that it now [leads the market in] Houston, Texas in services such as SEO, PPC and Social Media Marketing. I've upgraded lifestyles (by about a factor of 10) and have even been able to start a local charity organization called Hashtag Heroes (www.hashtagheroes.org), which helps local families in need by putting tangible items in their hands when they need it most!

Do you have any regrets?

I have ZERO regrets as life has never been better.

What's the best part of having left?

Having the freedom to work for myself whenever I want and wherever I want - sweet freedom. I also enjoy working with my wonderful team at SSD and how we all work close together in a family environment. I truly enjoy working with my crew every single day.

What surprised you most?

I was surprised that I would be able to succeed in such a down economy and double the business each year... I do not take this for granted, and I'm very thankful.

Do you think there is a type of person that is more suited to make this transition and if so, what do you think made them suited to follow their passion?

I'm not sure what the secret sauce to success is, but I feel that you need to be really good at time management, leading and building teams, watching your bottom line closely and always being ready to adapt and evolve your internal processes at a moment's notice. I also had the skill sets to do all of the work on my own for the first couple of years, which came in handy when I finally began hiring employees because I was able to closely monitor and manage their work.

Your Former Title: Not given by interviewee

Where were you formerly employed? A buying and marketing company in Houston in the following industries: security/CCTV, electronic MRO, electronic OEM, data communications, home automation and telecommunications

Please provide your current company or position: CEO of SSD, Inc. at FairMarketing.com

Interview: J-Coby Wayne

I left corporate America as soon as I worked up the nerve to do it so that I could work for myself and do business according to the principles and values that were important for me to stand for.

What prompted you to leave working for a corporation?

I had already built my own practice as an Internet in health care consultant, speaker, trainer and media figure. I realized very early on in my professional career that I didn't really fit a corporate profile. I found unethical and hyper-competitive behavior all around me, and I saw repeatedly that the least pleasant, least principled, most selfish, materialistic and over-achieving people at all costs were the ones who were consistently rewarded. I simply wasn't willing to do what was required to play the game. I literally had a performance evaluation consist of my supervisor telling me I had to start "wearing women's clothes, have a fancy haircut and wear more makeup" to get anywhere in the company. That's pretty much when I knew I was done with corporate. I did end up with a better corporate job with a guy who really rewarded me well, promoted me often and gave me many opportunities and a lot of latitude for creativity. But he got upset when I started getting a lot of high profile national media coverage, and he started making a huge amount of money off me, but not passing much share of that through to me. I didn't care about the money, but I did care about being valued properly for the value and visibility I was bringing to him and his firm.

I finally ended up at another health care web company where we agreed that my telecommuting from upstate New York was an experiment that we would try for at least a year. I could sense that towards the end of that year, they were thinking it wasn't working. I went on CBS News without telling them – which was not willful underhandedness or sneaking, but rather just politically ignorant – and they started getting lots of calls based on the interview. They weren't too pleased to be taken by surprise like that. Very shortly after I joined with them, I realized that they had absolutely no business plan except to be bought and that they lacked substance, so I was ready to move on. Rather than look for a corporate entity, I mustered up my courage and decided that I had enough visibility and credibility to take what seemed to be the huge risk to go into business for myself. It was the best professional decision I ever made, and I have never looked back.

How has it changed your life?

I have been my own scheduler and responsible for my own livelihood basically since 1996. I could and would never work in a physical office again. Even before I left corporate altogether, I always made it a stipulation of my being hired that I could telecommute. I was lucky to be an expert in a field at a time when there weren't many experts, so I was in a position to ask for whatever I wanted in any job I was considering. It created some resentment among others at the companies who never even thought to ask to telecommute.

Do you have any regrets?

Absolutely none.

What's the best part of having left?

It has given me the freedom to build an entirely principle-based enterprise that uses a completely new exchange model that is totally unlinked from fee-for-service commerce. I am able to make a difference by pioneering new ways of being and doing that are the antithesis of most corporate models. I am contributing to the evolution of enterprise, entrepreneurship and money, and I am able to help others do the same by moving beyond hyper-competition, beyond win-win to "grow-grow". I am part of the early wave of cooperative and collaborative-based enterprise now just as I was part of the early wave of the World Wide Web in the early 1990s.

What surprised you most?

How easy it was. My professional career exploded when I went into business for myself. Again, I was lucky to be only one expert in an entirely new and hot field just at the right time, so I was effortlessly in high demand. I wish I had left corporate not based on fear or being pushed to do it before potentially being fired.

Do you think there is a type of person that is more suited to make this transition and if so, what do you think made them suited to follow their passion?

I do really know this is not for everyone because I work with people and enterprises who want to be entrepreneurial, but are not at all suited to it. You have to be a self-starter. It sounds clichéd, but it's true. If you don't have the discipline, experience and excitement to work without any external stimulation or deadlines imposed by someone else, it's not for you.

If you don't like to be organized, it's not for you. If you can't work unless you are physically surrounded by other people, it's not for you. If you don't like having to deal with money and financials, it's either not for you, or you need to partner with someone who's good at that. If you wouldn't do what you do for free, it's not for you. Entrepreneurs and people who leave corporate do so because they stand very strongly for something they believe in and love. If you are risk-averse and are made uncomfortable by the cyclical nature of being in enterprise for yourself, it's not for you. It's not enough to have a good idea. It's not enough even to have loads of passion. You have to have the skills and will to execute – tirelessly – and the ability to wear EVERY hat all the time.

I also think you generally have to be outgoing, comfortable with who you are, social and community-minded. If you are more introverted, shy, inexperienced with being a friend, traumatized by being unpopular in the past or impatient or uncomfortable dealing with people, you probably won't do too well working for yourself – unless you can partner with people who really enjoy this so you can focus on holding your vision and creating ideas or inventions. Plenty of geeky, unpleasant, anti-social techies have been fabulously successful in business!

Your Former Title: Executive Vice President, Executive Editor

Where were you formerly employed? Harvard University, Institute for the Future, First Consulting Group, COR Healthcare Resources, *Medicine on the Net* Magazine, drkoop.com, onhealth.com, a consortium of international leadership training/consulting companies

Your current company or position: Co-Founder and Chief Experience Guide, Energy Arts Alliance, energyartsalliance.com; Founder, Poet, Visual Artist, e3 poetry initiative, jcobypoet.wordpress.com

Interview: Tom Ferguson

What prompted you to leave working for a corporation?

At the time, friends called it an "entrepreneurial seizure." I had worked at this company for 22 years and loved the business and the leadership – but, I guess, I loved more the idea of a new challenge in fulfilling the promises of self-help books and tapes that promise freedom in taking control of your life.

How has it changed your life?

For me, one objective was personal development ... to use God-given gifts to design mechanisms and artistry for client organizations to honor work and achievement.

Do you have any regrets?

None!

What's the best part of having left?

There is little more satisfying than to be recognized for helping clients as an outcome of your own creativity and passion for what you do.

What surprised you most?

Even with the world's best new mousetrap, people will not beat a path to your door!

Do you think there is a type of person that is more suited to make this transition and if so, what do you think made them suited to follow their passion?

Introvert/Extrovert … not so important as loving what you do and having the confidence and competence to characterize the value of your work to the prospect.

Your Former Title: VP Marketing and Business Development

Where were you formerly employed? ABB Randall

Your current company or position: CEO, Visible Applause, Inc., www.visibleapplause.com

Interview: Pam Terry

I was working as the Executive Director and President of the Galleria Chamber of Commerce from 1988 to 1991. Newly divorced and a young 37, I was loving my new life in a job that I loved more than any job I had ever had. I hit the ground running because it fit me so well on many levels. I did not have a college degree and had worked my way up the corporate ladder from being clerk at M.D. Anderson Hospital in the 70s to becoming what I considered to be a prestigious civic leader in Houston, Texas. I didn't realize till much later, but my father was so proud of me. And I felt I had arrived.

It wasn't the accomplishment of reaching this "pinnacle" of success that was the ultimate thrill for me, however. It was a position that I lived and breathed. I could use all of my talents, gifts and skills to make a difference for people. I loved working with groups of people and connecting businesses. I worked night and day. I was the second Executive Director for this young and vibrant chamber of commerce in a major metropolitan city. Little did I know about the politics that would choke me out and cause immense suffering!

As I successfully repaired broken relationships that the past Executive Director had caused, I was creating committees for member involvement. To this day, as far as I can tell, the Galleria Chamber still uses the structure that I created for member involvement. By the time I left the chamber, I had increased member retention from 25% to 75%, and there were over 100 members actively involved on committees.

During my first few months as Executive Director, I became accredited in public relations by the National Public Relations Society of America and I had completed the Landmark Forum of Landmark Education. Both of these efforts gave me incredible confidence. I began to see my talents, skills and flaws more clearly.

What prompted you to leave working for a corporation?

Nothing had prepared me for what was to come after three years when I was "ousted" by the Board of Directors. What happened? Well, let me make a long story short. We had an incredible sales person who was bringing in members left and right, and when she left, we hired a new person who simply was not a salesperson and new members dwindled. Every year, there was a new Chairman of the Board. After the third Board Chairman, my responsibilities remained the same, but my authority kept being limited. Little did I know how incredibly important it was to have clearly defined boundaries of authority. When the new members dwindled, so did income for the chamber, and I was blamed for it. Yet, I did not get to hire the person (that I wanted) for the sales position because the board wanted a selection committee to hire them. They picked the person, and I did not get to approve who they hired. The person I wanted to hire was not selected. The person I did NOT want to hire was selected. She was a PR person, not a salesperson. In the end, after three years, I was asked to leave, and it was difficult because my whole identity was wrapped up in being Executive Director and President.

How has it changed your life?

The whole experience had a huge impact on my life at the time. I struggled to "find" myself, and learned it's not about finding yourself. It's about creating your life the way you want it. One thing for sure, I am extremely wary of being the head of a non-profit. My eyes were opened quite a bit. Over the years, I have noticed that most chambers, the Galleria Chamber of Commerce included, do not keep the Executive Director/President for more than 3 to 4 years on average. Of course, there are exceptions, but generally this is the rule.

The good news is that I know what I am capable of, and I developed some great relationships. I succeeded in many ways in that job. I made a difference for people. I am glad that I did it. I saw an opportunity, I went after it. I had success, and I had failure. Now I am more of an expert. I saw what didn't work. Communication is key and so is understanding boundaries. I saw my weaknesses. All of the experience was extremely valuable.

After I left, I began to experience being an entrepreneur. I wasn't very successful as an entrepreneur, but I was beginning to develop a strong taste for doing my own thing and for developing respect for entrepreneurship. I began to hang out with entrepreneurs and started working with a couple of start-ups. I tried a lot of different things, including straight commission sales work and online marketing, and I began to dip my toe into technology. It was the beginning of my addiction to computers and the wonderful world of technology.

Do you have any regrets?

Yes, the only regret I have is that I wish I hadn't worked such long hours. Only because I wish I had spent more time with my son while I was working there. He was 5 to 8 years old during that time. He has since told me that he was glad I left there because I was always working. I didn't realize the impact that it had on him.

What's the best part of having left?

The best part of leaving is that it really set me free to create my life the way I want to on my terms. I started working in the technology field in sales for the next 15 years and worked my way up to being Vice President of Marketing for an information technology consulting company before I was laid off. Since then I have been working as a sole proprietor for the past 4 years.

All of my previous experience in technology, marketing, operations, public relations, non-profits and membership organizations serves as a strong foundation for what I have to offer and utilize.

What surprised you most?

What surprised me the most about leaving the chamber of commerce was how much my identity was wrapped up in that position. I was devastated about leaving, but in the end, it was the best thing for me because I found that my most important value is to feel free.

Do you think there is a type of person that is more suited to make this transition and if so, what do you think made them suited to follow their passion?

Being an entrepreneur takes passion and guts. But, mostly it takes passion. Some people are completely happy to have a job for whatever reason. And, we need people to be in jobs. To follow your passion takes courage, and wisdom helps a lot. I would suggest that to follow your passion, you should become part of a community that supports you. I have found that being a member of Powerful Women International has been one of the best communities for supporting me to live the life of my dreams.

Your Former Title: Executive Director and President

Where were you formerly employed? The Galleria Chamber of Commerce

Your current company or position: Public Speaker Coach and Communications Trainer; Chief Operating Officer, Powerful Women International

In our first chapter, we explored the evolution of one particular sales shark into recovering sales shark, and we shared the stories of other successful people who left the shark tank.

But what's the connection to social media? Well, social media has completely changed the sales and marketing environment. It's not the same old sea where sales sharks rule with their "eat what you kill" approach to business. The market, marketing and sales have evolved – thank goodness!

In this chapter, we'll explore exactly what this evolution looks like in order to make the case for why today's market creates a unique opportunity for sales sharks to change and why it's so important to understand and use social media effectively.

Now, let me state here that I have tremendous respect for the sales field and sales professionals. Many great, talented people are attracted to work in sales. Sales professionals are some of the highest achievers I know. And corporations bring so much value to the world by generating wealth and opportunity for millions of people.

But how much more would sales professionals and corporations be able to contribute to their communities and societies if they shed that sales shark persona and looked beyond the bottom line with the Give to Grow philosophy, which recognizes that doing good automatically means you do well?

Can you imagine a world where all sales professionals and corporations shared information freely instead of hoarding information as an advantage over the sales guy down the hall or the corporation down the block? Can you imagine a world where all those high-octane sales sharks and corporations turned their full attention and creativity to how to do good for EVERYONE with money exchange and profit coming as a natural by-product and result of acting from goodwill and supporting the growth of all – even one's "competitors?"

It's starting to happen, and social media has been a critical driver in changing the nature of the "business sea." Social media changes what's "rewarded" and the talents and skills that are valued. Let's explore how.

The Evolution of Salesmanship: From Competition to "Co-opertition" to Cooperation

In today's world of social media communication, sales sharks just don't thrive. Coming from an attitude of "kill or be killed," "eat what you kill," "take no prisoners" is very out-of-date thinking and doesn't work when using social media to build brand awareness, engagement and community. Why this doesn't work is that social media gives us the ability to be totally transparent and give to grow.

On social media, we MUST come from a place of adding value for our clients, prospects, followers, friends, connections and fans. Social media users are savvy when it comes to knowing what is authentic vs. pure sales. The way we meet their needs is by providing information that they find valuable - value according to them, not according to you!

As an expert in your field, share content that is requested by your community of readers, followers and viewers. If you don't know what they consider valuable, ask them! Simple polling sites like www.surveymonkey.com work great for getting input from your audience.

With the ability to easily read someone's digital body language, now is the time to consider how you are being interpreted when you post on your social networks and through your email marketing newsletter. If there is no value in it for the recipient, don't send it. Always ask yourself, What's In It for them (WIIFT)?

The great news is that our evolution has brought us to a place of cooperation where we share what is important with the masses via social media, and we do not worry about people taking what we have or using what we have. The very nature of social media is that it helps people and businesses foster trust and loyalty.

It is very hard to give up the idea of needing to protect what is ours because other people will take it. I came from "scarcity" all of my life. It was a life of competition and one-upmanship. It was not a feel-good place at all.

With the advent of social media platforms, it has become apparent that this is a land of cooperation. Of sharing, re-posting, forwarding and exchanging blogs, articles, others posts and pictures and philosophies and ideas...Sales sharks need not apply. Wow, what a difference! Going from a scarcity mentality, which is what most sales shark dinosaurs come from, to a place of abundance truly changed my world and can change yours, too.

Springboard Tip
Practice makes perfect. When you find yourself losing energy, tightening up and not feeling at peace, you are not coming from abundance. Breathe, ask to see the situation differently and change the lenses of your glasses!

When I was a young salesperson, I always questioned what I didn't have. I always compared myself to everyone else. This is a huge mistake. First, it's a total waste of your life blood, your energy; secondly, there is only one you and no one else can be you; third, people will read your body language, be it digital or in person, and will not want to do business with you because they sense this "lack" that you have made up.

For the same reason we naturally gravitate towards positive, upbeat, happy, confident people. We do not want to be around people who think they are not enough or are less than. We can read this mindset and smell it. It is a huge repellent. Just as a dog senses fear in a person, we as humans have an innate ability to sense where we each are coming from.

Giving up your sales shark's love of competition and scorch-the-earth mentality will be a shift. It will take practice. The great news is that there is enough for everyone. You will attract who you are supposed to work with, and just as our Earth has magnetic fields, you will resonate with the vibes and bring prospects and clients into your orbit. As we learned in Chapter 1, having trust is a big part of that.

In the spirit of best practices when incorporating social media into your marketing plans, selling just doesn't work. Period. I would tell you to experiment and give it a try, but you may only have one chance to make a first impression. The best advice I can give you is to always come from giving and adding value. It's okay to offer up specials and discounts and deals, just do it no more than 20% of the time when using social media tools.

The bottom line is this: It's all about adding value, giving to grow, helping others when you can and making a difference. Building trust is crucial to success in business and in life. Trust comes from adding value, from cooperating and not being fearful of losing anything. Once you have that trust, you have an opportunity to have a customer for life.

So, shed that sales shark... If you don't, you are a dinosaur waiting for a tar pit!!

Does Anybody Really Like to be Sold Anything?
Or Enrolling vs. Selling

Does anyone really want to be sold anything? We want to know features and facts, but we want it to be from someone who believes in the product and is passionate about what they are selling. If you are passionate and believe in what you are offering, you really don't have to focus on closing the deal or selling the deal.

If you can touch, move and inspire people with your passion and purpose, you are enrolling them. Being enrolled in a product or service or idea isn't selling. You are coming from a place of believing, in having conviction around this vs. selling.

We all want to be recognized, acknowledged. It's part of the human condition. So how do we do that online to show we care and are listening? It's easy. Always respond to invitations to connect or requests to like with acknowledgement. Thank people, write a sentence or two, show you care! This says tons about your digital presence. And it differentiates you from just about everyone else out there.

How you respond to a new fan on Facebook or a request to connect on LinkedIn is part of your digital body language impression that you put out there, and its energy circulates in the blogosphere.

Let's use LinkedIn as an example. Out of the box, LinkedIn gives a default canned statement when you send out a request for someone to connect with you. Rather than just send the "canned" statement, take 20 seconds, and customize it with a personal note about why you are requesting to connect. People want to know you care. People want to know you are "listening." This simple action tells about who you are as a person.

What message are you sending? Are you caring, too busy, thoughtful, thoughtless, distracted, self-important...? You get the picture. It only takes a few moments to jot a note about why you want to connect or how you know each other. It shows thought and personalization.

Some of us are better at giving vs. taking. We tend to personalize and care naturally, which is pretty much opposite the natural behavior of a sales shark. For sales sharks, it's generally easier to ask for and take what you want.

Qualities of the Sales Shark	Qualities of the Relationship Builder
• Greedy	• Altruistic
• Selfish	• Selfless
• Self-centered	• Aware of others' needs
• Egotistical	• Thoughtful
• Take no prisoners	• Cooperative
• Hungry	• Abundance-minded
• Me first	• Helpful
• Give to get	• Give to grow
• Talks more than listens	• Listens
• Thinking ahead of the present moment	• Stays present
	• Collaborative
• Hyper-competitive	• Authentic
• Ruthless	• Patient
• Chameleon	• Trustworthy
• Shyster	• Builds relationships
• Hustler	• Honest
• Pushy	• Engages
• Dishonest	• Has integrity

Source: Photobucket

Source: Photobucket

This particular recovering sales shark called Shelley finds it is much easier to give than take. Just as the tides ebb and flow, you must be conscious of what you give, and know you have a right to receive back. It's the flow of the world!

When communicating inside of social media tools and email, it is very apparent when someone makes it personal. But sometimes being appropriately personal can be hard since business is business, and some of us like to draw a thick line. Social media tools are blurring the lines between business and personal. Facebook, developed for a network of college kids and their personal lives, has now morphed into a business tool. Why? Because people buy from people they relate to and like. It's about relationship marketing – "REAL-ationship Marketing." (See my other book, *GET REAL, People!* for a more in-depth discussion of REAL-ationship Marketing.)

For most of us, our business audience is on Facebook. This being the case, find out a tidbit about a person AND make a more PERSONAL connection with them, like recognizing their birthday or finding out what they like, who they relate to, where they hang out. With that, you have the first step in being real and getting "related" to that person. I can't repeat it often enough: People buy from people they relate to and are engaged with.

As a sales professional, being enrolled in a product or a service or idea is very different than selling. You are coming from a place of belief in yourself and what you are offering. I learned this from an education company that never, ever spent one nickel on marketing or advertising. It would amaze me that all their classes were filled and I wondered, How did they get so many people to attend without marketing dollars?

It turns out that those who attended their classes received so much value and growth that they went out and evangelized and preached the gospel on how these classes made a huge difference in their personal and business lives. They walked the walk and talked the talk. They were "being" the change they wanted to see in the world and just from that, they enrolled people in the program effortlessly. People wanted to drink from the fountain of their passion.

Engagement Marketing:
The New "Non-Sales" Technique

When we talk about shedding the sales shark, we must ask ourselves what to replace our behavior with. This chapter is all about the new selling or "non-selling" in the traditional sense. It is about relationship marketing. Building relationships. Engaging with our listeners. Listening instead of speaking. Connecting, conversing, building trust.

During my years in business, I've discovered that 90% of business comes from current customers' referrals and 10% from new customers. With that in mind, ask yourself how much time you spend on really engaging with your community of people. Social media has revolutionized the way we meet and communicate, and it gives us the greatest platform for growing relationships through giving to grow.

Engagement marketing is all about word-of-mouth marketing. The vast majority of our customers come from our other customers referring business to us. That's a large part of why we engage and offer value to our listeners on the various social media platforms. You can't buy word of mouth, and you can't sell word of mouth. When we provide value and a great customer experience, new customers will come to us. You will never have to sell again.

Social media gives us an opportunity to participate in discussions with our fans and friends. We can sit back and lurk and read to our heart's content.

But it's in the participation, in the engagement, that the magic happens. And in this day and age, in order to fully engage, all businesses that have a physical door also need to have a "cyber-door" and give people a virtual experience because that is one of the main ways your customers and prospects engage with others.

In my experience, buyer behavior shows that we trust comments and reviews even from strangers before we trust marketing via traditional means. Before we travel, dine and make purchases, we ask our friends and research what others think of products we buy and the people we deal with. Social media gives us the opportunity to spread endorsements far and wide.

I find it amazing when I look at the analytics of my various social media platforms and see the far and wide reach of social media. It's always a pleasant surprise when I hear a thank you from someone in another country or just in another state in the United States. Your message goes far and wide, and social media gives you a great platform for building relationships and engagement, so selling becomes a thing of the past!

Your customers' friends, fans, connections and followers are your most valuable prospects. Engagement drives social visibility, which drives growth.

Springboard Tip
In engagement marketing, make it easy for people to connect with you. Instead of saying, "Find us on Facebook," give them the exact Web address on your print materials, in your videos, even in person.

Relying on traditional marketing approaches like blasting out direct mail pieces or unsolicited email messages to the masses - or as I call it, "scattershot marketing" - just doesn't work anymore. With social media marketing, the message you send out is managed by your network of friends, fans, followers and connections. They are the ones who will influence your customers and prospects. The new world of engagement marketing allows us to participate in the conversations and discussions that are occurring online 24 x 7 x 365.

With engagement marketing, it is extremely important to listen to our social media community. Everyone has a choice in who they want to follow and engage with, so when someone reaches out to you, it's imperative that you respond and let them know you hear them in a timely fashion. Prospective purchasers of your products and services want to be treated as participants in the process. They want to do business with companies that they know and trust and communicate with long before they need your product. Establishing a community of engaged fans and connections is crucial to your success in this new marketing and non-sales arena.

Now, there's no doubt this takes work. If you don't stay on top of listening, connecting, communicating and engaging regularly and in a real and genuine way, you can fall into the trap of "social negligence" (a concept I heard in a podcast by B.J. Emerson) – which means not listening and responding to your community.

It is all about creating a dialogue between your organization and the people in the markets you serve. People will voice their opinion about your brand and may even become brand evangelists for you online. How sweet is that?! Your super-fans will "sell" for you. They will expound on the virtues and the values of your company. And, when this occurs, you can kiss having to work at sales, make cold calls and hit quotas goodbye!

> **Springboard Tip**
>
> An introduction through a friend carries the power of trust… and that endorsement is better than any marketing list you can buy!

Online engagement marketing serves customers, prospects and organizations, delivering tangible benefits to all. The more engaged a company is with its marketplace, the more successful it will be using social media tools, and the further it will distance itself from traditional sales shark techniques of one-way communication and monologue. Through social media, we are now able to have dialogues!

Getting Real with Social Media

The good news about social media is that being real is rewarded there.

What social media did for me was something I never thought would happen. It gave me a voice where authenticity shines through. Where real is rewarded. Where giving to grow is an example to share. By being fired, I had to go forward fearlessly and not let anything hold me back.

Social media is a natural way to help other people shed whatever is holding them back from being real. For me, it was the sales shark. I didn't have to sell; I had to serve! The more I serve, which is my passion and my nature, the more I am rewarded. Social media is just an incredible medium to do that.

With social media, it's all about "listening" to what people are saying and giving to grow. Being transparent is what people relate to on social media. Being transparent online or offline is something that is powerful but not easily measurable in terms of traditional sales and business metrics. Social media tells us loud and clear in the users' own words and behaviors that transparency is what people want. They want to see you with all your weaknesses as well as strengths (being Flawsome). Contrary to what many people find in corporate settings, social media's culture is all about acceptance, inclusion, sharing and celebration – from social causes and bettering the world to the hobbies and things we care about and love to do personally.

Here are a few best practices for engagement marketing with social media:

1. Conduct regular polls and surveys of your database to ensure you understand the current needs and wants of your market.

2. Strive to integrate customer feedback as much as possible in order to improve your products and services.

3. Understand the power of social media and have active, current profiles set up on all the popular social sites such as Facebook, Twitter, LinkedIn, and Google+. Focus on the one channel where the majority of your customers are.

4. Have effective listening and monitoring systems in place on all social media.

5. Have a corporate social media policy in place that lets staff know what can and cannot be said, what actions can and cannot be taken and how to handle any negative situation.

6. Use a reliable customer relationship management system.

7. Conduct regular training sessions for all members of staff on proper customer relations and social media best practices.

8. Stay on the cutting edge by evolving, adapting and integrating new technologies.

9. Embrace high-tech but always maintain high-touch by reaching out to your customers, prospects, vendors and partners.

10. Have a very high customer satisfaction rate.

11. Consistently go out of your way to let your customers know how much you value them.

12. Drive relationships, what I call "REAL-ationships," by providing added value via news, blogs, entertainment, engagement marketing. OPC – "Other People's Content" – is readily available. Remember, it's about adding value and giving credit, not taking credit.

13. Always ask your connections for input. They want to be part of the community and it's important to give them an opportunity to participate.

By studying and integrating each of these best practices, you'll go a long way towards improving your success through engagement/ relationship marketing. Your customers will not only like you, but they'll love you!

CHAPTER 3
Social Soup: A Guide to Social Media

Now that we've explored how social media is changing the market and sending the sales shark the way of the dinosaur, let's tour each of the major social media outlets and learn some practical tips for using them effectively.

Social media has been a great springboard for me to get real and apply my natural-born sales skills in a more enlightened and less stressful way. It's put me in touch with so many people who are interested in growing their business. I'm always glad to see that in discovering how to use social media to grow their business, many of my clients also end up growing themselves as people and learning the same lessons I've learned.

One of the reasons my business and I have been able to thrive through social media is because I've listened to questions that come up again and again. I've created my workshops and other services in response to these specific questions.

If you are wondering how to use social media – or how to use it better – you have probably asked yourself the same questions.

- What is the best recipe for social media?
- How much should I do?
- How much content should I post?
- How many times do I post?
- When should I post?
- Which are the best social networks to use to communicate?
- Where do I find content to post?

This chapter addresses all these questions with a recipe book to help you cook up great ways to expand and engage with your community.

As you consider using social media to grow yourself and your business, just remember it's all about adding value for others... Social media gives you a means to know, share and be who you really are.

People want to do their homework when ready to buy and then seek out a person to buy from whom they trust. Social media has leveled the playing field for all business owners, from solopreneurs to *Fortune* 100 companies. Social media gives you a chance to share you and your personality with the world and to ditch that sales shark. IF you give value, you will grow. If you care about others, you will grow. This is the Law of Abundant Exchange.

This chapter on Social Soup gives you an overview of the top social networks and how to set them up and use them to grow your online presence and set you and your business free from having to focus on selling.

First let me define social media:

> *Social media consists of web and mobile applications that allow for two-way communication and interaction with the exchange of user generated content.*
>
> *Social + media = social media.*

The common thread is the blending of technology and social interaction to create value.

The top social media platforms are Facebook, LinkedIn and Twitter, plus YouTube, Pinterest, Google+ and a variety of blogs, podcasts, video-sharing sites and social bookmarking. I'm only going to cover the social media networks that I think are the most important and here to stay.

Here are some important facts about social media from Wikipedia and other sources:

- According to comScore, social networking accounted for 22% of all time spent online in the United States in 2011. In the same year, social networking sites reached 82% of the world's online population.

- According to mobiThinking, there were more than 5.9 billion mobile device subscribers globally in 2011.

- According to ctia.org, a total of 310 million people in the United States used mobile devices as of June 2012.

- According to Twitter, there were 140 registered Twitter users in March 2012. According to Media Bistro, as of June 2012, there were an average of 400 million tweets per day.

- According to the Pew Internet and American Life Project, as of August 2012, 34% of seniors in the US use social media sites, 18% of them on a daily basis.

- As of October 2012, Facebook has **one BILLION users**... and counting.

- Facebook tops Google for weekly traffic in the United States.

- Social media has overtaken pornography as the #1 activity on the web.

- iPod application downloads hit 1 billion in 9 months.

- If Facebook were a country, it would be the world's third largest.

- 93% of businesses use social media for marketing.

- A U.S. Department of Education study revealed that online students out-performed those receiving face-to-face instruction.

- YouTube is the second largest search engine in the world.

- In four minutes and 26 seconds, 100+ hours of video will be uploaded to YouTube.

According to a report by Nielson:

> "In the U.S. alone, total minutes spent on social networking sites have increased 83 percent year-over-year. In fact, total minutes spent on Facebook increased nearly 700 percent year-over-year, growing from 1.7 billion minutes in April 2008 to 13.9 billion in April 2009, making it the No. 1 social networking site for the month."

The main increase in social media has been on Facebook. It has been ranked as the number one social networking site. Approximately 50% access this site through their mobile phone. In an effort to supplant Facebook's dominance, Google launched Google+ in the summer of 2011.

With this understanding of social media's growth, scope and importance, let's look at each of the social networks I think are the most important to use to set you and your business free of selling – online and in life!

Facebook: The Main Ingredient

If there were ever a network that influenced our behavior and set the standard on social media for business, Facebook would be the one. With over 1 billion members, Facebook is the king of social media. It's easy to fall into the sales shark mode when using Facebook and come from a place of sell, sell, sell. However, this will not contribute to your success on Facebook.

From my experience over the last six years, the best way to grow is to be who you are and share that personality on all social networks. You will not need to sell if you Give to Grow. Whether you are giving tips on business or philosophical tips, making people laugh or encouraging participation, it is always about connecting and engaging with your community.

What follows is pretty much technical how-to information from Facebook, but in my opinion, the real secret to success on Facebook is having fun and engaging with your fans in your own unique way.

According to Facebook, over 1 billion people like and comment an average of 3.2 billion times every day on Facebook. When you have a strong presence on Facebook, your business is part of these conversations and has access to the most powerful kind of word-of-mouth marketing — **recommendations between friends**.

Facebook recommends five steps for using Facebook for your business:

1. **Create a Facebook Timeline Business Page:** It's free to set up a Page and it only takes a few minutes to get started. Go to www.fb.com/pages to do this.

   Take your time and look around at all the pages similar to yours. See what you like and what you would modify.

Here's what my Facebook Timeline Business Page looks like:

Choose a Category and a Page Name that represents your business. Include key words in your page name.

Pick a logo or another image that people associate with your business to use as a profile picture.

Write a sentence about your business so people understand what you do. Include key words.

Set a memorable web address (URL) for your Page that you can use on marketing material to promote your presence on Facebook. www.facebook.com/username.

Choose a cover photo that represents your brand and showcases your product or service. It's the first thing people will see when they visit your Timeline Page. (Make sure you read Facebook's terms and conditions on what you can and cannot do with your cover photo at www.facebook.com/help/pages/new-design.)

2. **Connect With People:** Now that you have a Business Timeline Page, it's time to reach your current and potential customers. Connect with people via your personal Facebook page friends by inviting them to like your new Facebook Business Page. You also can import your contact list from Outlook or Apple, upload your email service provider contact list and find your contacts in your Web-based mail service such as Yahoo or Gmail. All of these will help you grow the number of LIKES on your Facebook Business Page organically.

3. **Engage Your Audience:** When you post content and have people liking, commenting and sharing on your Page, you're building loyalty and creating opportunities to stay top of mind with your fans on Facebook. Keep in mind a good rule of thumb is 80% educational, entertaining and engaging and 20% business related on what you are posting. Experiment and learn how to create content that will keep your audience interested. If you can stay top of mind with your audience, when they need what you have, they will think about you! Remember to always ask before you post, "WIIFT: What's In It For Them?!"

4. **Post Quality Content Regularly:** When people like your Page, they're saying that they care about your business and want to know what's going on. They have opted into hearing from you. Posting relevant content is the most important thing you can do to keep them interested.

Some ideas on writing quality posts include:

- Make sure your posts are relevant to your audience and business.
- Share bits of your personality.
- Be succinct, friendly and conversational.
- Share photos and videos. They are the most engaging content.
- Ask questions or seek input: KISS-Keep it Short and Simple.
- Give access to exclusive information and specials.
- Be timely by posting about current events, holidays or news.
- Make it easy for people to participate in your posts by asking True/False, Right/Wrong and A/B/C questions.

Springboard Tip
Post at least 1 to 2 times per day so you stay top-of-mind and relevant to the people who like your page.

5. **Influence Friends of Fans:** When people interact with your Business Page, their friends can see it in their news feed. Encourage people to interact with your business. We've said repeatedly that word of mouth is the strongest form of advertising. When someone interacts with your business on Facebook, it creates a story. People can see when their friends endorse your business by liking your Page or connecting with it, and it can influence their own decisions.

Here are some ways to encourage people to interact with your business:

- Encourage people to LIKE your Business Page with a sign in your store or by offering a special discount to people who check in via Facebook Places.

- Create events on your Page, and invite people to join them.

- Ask questions, and create posts that encourage engagement.

- Share exclusive information and deals that people are likely to want to pass along to their friends.

- Be engaging, and ask people to participate with calls to action with each of your status updates.

80% of consumers say they are more likely to try new things based on a suggestion by a friend in social media.

- Facebook changed the way users interact with content on Business/Fan Pages: **Any user, including non-fans, can now post on Fan Page walls** and **like/comment/share Fan Page content**. In other words, a Facebook user does not have to first like your Fan Page before they can interact with your content.

- The emphasis has shifted from gaining more likes/fans to **increasing the number of likes, shares and comments on each piece of content**. Shares include: all comments, likes of comments, sharing comments, sharing photos, sharing updates, sharing videos, posting events and sharing events. Facebook uses a proprietary formula called Edgerank, which determines how many will see your content when you post.

> **Talking About This**
>
> The number of unique people who have created a story from your Page post. Stories are created when someone likes, comments on or shares your post; answers a question you posted; or responds to your event. Click on the number to see more details. Figures are for the first 28 days after a post's publication only. Click on "Talking About This" to sort your posts.

- The average page only has a 2-5% "talking about this" statistic based on Facebook's Edgerank. The burden lies with the page creator to make the content they share engaging so more people will see it and pass it on to others.

- Create your Business Page updates in a manner that will naturally inspire fans and their friends and visitors to your page to share with their networks. The goal is to **set in motion ripples of viral visibility**. Basically, free additional exposure.

Top Tips for Getting More
Likes, Comments and Shares on Facebook

1. Post an eye-catching image/photo. It can be humorous. People love pets, kids and cute stuff.

2. Keep the narrative short.
   - A study by Buddy Media showed that *posts 80 characters or less in length receive 27% higher engagement rates.*

3. Do not write in first person all the time.
   - Write in a way that could sound like it was coming from others. Here's an example: "How interesting is this?!" It could be said by anyone.
   - You can always use first person in your comments when responding to bring in the more personal touch.

4. Include a call to action.
   - Simple calls to action such as "Click like if you agree" often work well. Asking people to add their comments is good too. But, it's the Share that will likely create the greatest exposure for your page/profile. So, ask them to share.

5. Write about timely topics with helpful tips and resources.
    - Provide plenty of information to share up-to-the-minute tips, new strategies and useful resources on your Facebook Page and/or profile. The bottom line is for your community to find extreme value in reading and sharing the post.

6. Take time to thank those who share your content. People love to be acknowledged. Let them know you care and are listening. Booshaka is a cool app that lets you see your top fans so you can acknowledge them.

7. People love to share Causes. It's the emotion that catches them and tells their heart (not their mind) to share.

LinkedIn for a Great Appearance

What is LinkedIn?

So many people come up and ask me, "I'm on LinkedIn, now what?" that I actually created a course with that title.

LinkedIn is a misunderstood social network, mainly because people think it's just for finding a job or being found for a job. So let's see the facts that contribute to this misunderstanding: According to The Reppler Effect blog, 93% of recruiters do use LinkedIn to search for candidates and research prospective candidates, as do hiring managers looking to find talent directly. But, our discussion in this book is about using LinkedIn for business development.

Let's start with a definition. LinkedIn is a social network that has over 185 Million users. It encourages you to network professionally, post and find jobs, ask and answer questions, join and create groups and build brand recognition - all while growing a network of connections.

You can research businesses and discover who you are connected to or find people you want to meet at various companies. LinkedIn also lets you feature your business via a company page and generate company followers, likes and reviews.

What makes LinkedIn different from other social networking sites is the buttoned-down feel of it. You won't find any games being played. It's fairly formal with not a lot of personal sharing.

LinkedIn is ideal for sharing business ideas, projects, licenses, white papers and anything else that helps put you and your business on the map. Keep in mind that LinkedIn is searched by all search engines, which helps your SEO (search engine optimization) as a side benefit.

Before jumping into using LinkedIn as a social network for your business, you must first create your personal profile as a businessperson. So, start by completing your profile page. The one thing you do not want when starting off a prospective relationship on LinkedIn is to have an incomplete profile. The bar that shows % of completion should be at 100%. As of this writing, LinkedIn is changing the algorithm on how you can a get 100% complete profile. One of the new items you need now is listing three of your skills and having your connections endorse those skills via recommendations from people on LinkedIn – which is an ideal opportunity to put give to grow into action! Give two or three recommendations and endorse other people's skills, and I bet you have people returning the favor. Another new hurdle is that you must have a minimum of 50 connections and also list two past positions.

Here's how to get going on LinkedIn:

1. **Set up your Profile Page**. NOTE: Turn off your profile update feature prior to editing your page. Otherwise, your connections will receive updates from you each time you make a change! Go to Settings; Profile; Turn on/off activity broadcasts. See below:

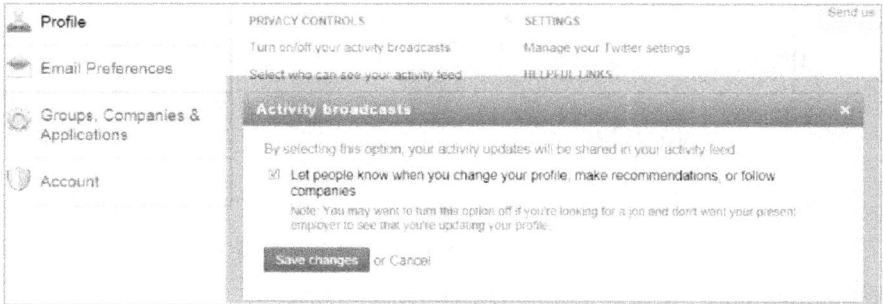

2. **Upload an Image**: You want to include your full name and a professional image. Research shows that you are 40% more likely to be found with a picture. I know I never connect with any request if they do not have a picture. Invest a few dollars for a professional head shot. Your picture speaks a thousand words!

3. **Personalize your Headline** or as I call it, your "bumper sticker." Most folks make the mistake of putting their title under their name. This is not how you will be found when people are seeking your services or products. Make sure your bumper sticker tells what you do and has key words that are searchable.

4. **Add History, past and present:** I believe you should share your work history, as well as a few personal nuggets about other activities in your life that add color and personality to your profile. For example, if you are a little league coach or sing in a choir, why not add this to your current or past history? This gives viewers a snapshot of you and may help them identify with you. I have received more comments on

sharing this type of information than I ever have on where I used to work.

5. **Create your customized URL.** Under the Edit tab, look for "Public Profile" and from there, select Edit. You will see a prompt to create your customized URL that will look like this: <u>linkedin.com/ in/shelleyroth</u>. (You will use your name or a variation of your name if it isn't available.)

Your public profile URL

Your current URL
http://www.linked.in/in/shelleyroth
Customize your public profile URL ▪ View your public profile

---

6. **Create Custom URLs for websites, social sites.** Out of the box, LinkedIn has websites listed as: Company Website, Blog, etc. Go in and choose "other" to add your own title and give it some marketing juice. So, instead of Website, you might add, "Visit our Website for Tips on Marketing." Do this for each piece of real estate you have. Three URLs can be added. You may have one link to your Facebook Business page. Remember, make it easy for people to find you and learn about you.

Edit Profile   View Profile

Additional Information

Websites:	Other	▼	Shelley Roth Landing Pa;	http://www.shelleyroth.co	Clear
	Other	▼	Videos on Using Social M	http://www.youtube.com/s	Clear
	Other	▼	LIKE on Facebook	http://www.facebook.com	Clear

---

7. **Sending Requests for Recommendations and Connections.** When you are asking for recommendations or for people to connect, do not

simply use the canned message that LinkedIn creates for you. Take five seconds and create a personal message that connects you to the recipient. This goes a long way in getting the relationship off to a great start.

8. **Add your Skills to your Profile Page at <u>www.linked.com/skills</u>**. A brand-new feature of the personal profile is the ability to add skills and have your connections endorse them. Go to the Skills section on LinkedIn, and add them to your profile. You will notice that LinkedIn will prompt you to "endorse" other people's skills, and they in turn will be prompted to do the same for you.

9. **Complete your Summary Page.** This is your opportunity to be engaging with your viewers. Tell a story, in the first person, about you. Make it interesting. You have just a few seconds to make them want to read on and learn about you. Add key words in your summary. Remember, your profile is searched by Google and other search engines.

10. **Grow your Connections.** Upload your contact list to LinkedIn, and then invite those you want to connect with to join your network. Remember: Include a personal note with each request. Another way to grow your connections organically is to join your college alumni group or special interest groups, etc., and reach out to others by participating in group discussions. Be sincere when you join these groups. You do not what to be just a "lurking" sales shark in the group, but actually participate. You also definitely don't want to sell inside these groups. Growing your connections on any social network is about like-minded people coming together and sharing information. It's never about selling!

NOTE: At the time of this book's publishing, LinkedIn was updating the look and feel of the personal profile page.

Company Pages on LinkedIn - Build It

So many of my clients don't even realize they can have their company page on LinkedIn, which can feature their products and services for free. If you don't have a Company Page, go to www.linkedin.com/company/add/show to add one.

Your employees are on LinkedIn, both current and past, so have your company listed as part of their profile page. Why not give the company more reach by building this page? Every person that connects with employees will also have the ability to go to the company page and learn about your products and services. They also can choose to follow your company, which means they will receive all of your company status updates. In addition, anyone browsing on LinkedIn for what you have to offer will find your company profile.

You can feature products and services and users can recommend them inside of your company page. A very good example of this is Dell Computer. Check out how they feature their products and services and the recommendations received.

Okay, so now you have your products and services on your LinkedIn Company page, now consider the product awareness that is made available to the LinkedIn community. LinkedIn lets you narrow the field to the products that you want to stand out. Prospective users also see which of their connections are recommending the product or service. This form of product awareness is hard to find and uses social networking in its purest sense.

Here's how to build your company page:

1. **Include all your basic company information and a great cover photo banner:** Include descriptive information that informs the reader of

what exactly your company does. You also can post jobs under the tab called Careers, although LinkedIn charges for this. What better way to find talent than on LinkedIn via employees and connections? Below is a screen shot of Springboard's company page.

2. **Include Products and Services.** Talk about a great way to advertise for free with banners and videos! Go to the Products tab on your company page and click Admin tools. Here you can add products and services.   LinkedIn provides easy-to-follow, step-by-step instructions to help broadcast those services.

3. **Promotion via Ad Campaigns.** If you have the budget, you can run an ad campaign, which LinkedIn walks you through. Even if you do not use this feature, I suggest you create the campaign just to view demographics on LinkedIn that would fit your target market. (Don't

worry: You don't have to pay until you reach the end of the campaign creation. Go to **www.linkedin.com/advertising?src=en-all-el-li-hb_ft_ads&trk=hb_ft_ads** for Ad Campaign steps.

4. **Measure via Analytics.** As the Company Page administrator, you will be able to access analytics about your page. You will want to monitor what is working and what is not. You can track page views and unique visitors to your page and tabs on your page. You can see how many clicks occurred for your product or service and how many members are following your company for given periods of time. You also will be able to view your page followers' insights.

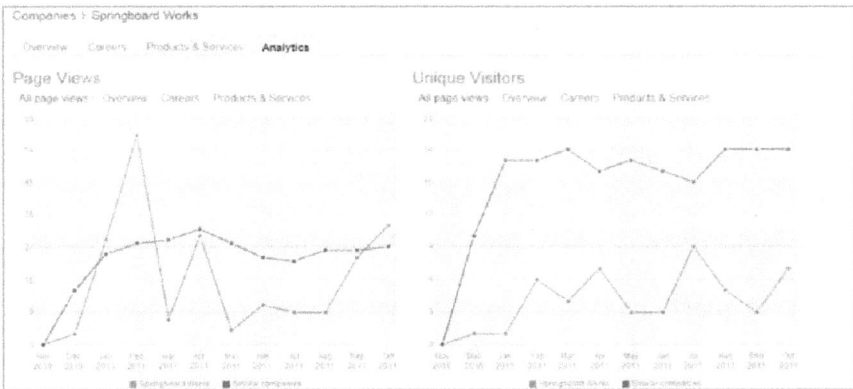

Company Status Updates on LinkedIn

With the launch of the Company Status Update feature for Pages, selected administrators can update status and share information on what's going on in the company. This gives the company an opportunity to share and be more personable with followers. The updates appear on followers' LinkedIn home pages, as well as the company's page itself. LinkedIn users can "like," comment on and share a company's updates, which is very similar to Facebook.

How to Be Found on LinkedIn:
Generating Interest for What You Have to Offer

Okay. So you have your personal profile ready to go. And your company page is set up with products and services for people to recommend. Now what?

Well, you certainly don't want to "sell" using social media. Sales sharks, chest thumpers and beating your own drum rarely attract people to you. What you want to do is Give to Grow. What I mean by that is there are so many avenues on LinkedIn to provide value to your audience. You can join various groups and contribute to the conversation by sharing your expertise, adding links to blogs, videos, information that is useful and by offering advice, industry insights or general information. You can do this not only inside of the groups you belong to, but also in the rarely used Question and Answer section of LinkedIn. By answering questions, you can establish yourself as a brand expert and meet many new connections.

By asking questions that engage and contribute, you will begin to connect with people who may be prospects, colleagues and potential customers.

This is all about giving to grow. Once you have connected with your community and add value, then it will be automatic that they seek you out when they need what you have. They may sign up for your newsletter or be directed to your website or Facebook page, but the bottom line is, add value first!

All the while, remember to be your REAL self, not the person you think they want you to be, but your real, loveable, knowledgeable self....filled with expertise.

Generating Interest via LinkedIn

Here are some practical tips for generating interest via LinkedIn:

1. **LinkedIn Question and Answer**: This is one of the most useful and underutilized tools on LinkedIn. This is the place I go when I cannot find answers on a Google search. It is a place to establish yourself as a brand expert, with knowledge and advice sought out by others.

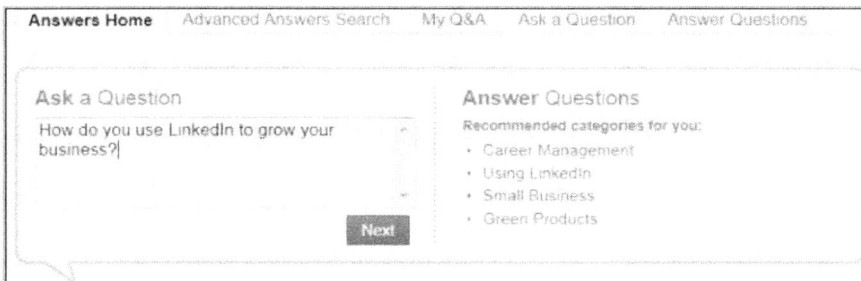

2. **Use Applications** available on LinkedIn to add value to your connections. You can bring in blog posts or let people know the business books you are reading and set up polls using LinkedIn. What a great way to create a focus group via polls for finding out what your connections are interested in!

3. **Join Groups** and participate in the conversation. These groups can be like-minded business interest groups, college alumni and past company alumni groups, and also special interest groups. You also may want to start your own group, but remember, it takes time to administer the group, so do your research on how to run a successful group on LinkedIn, and ask successful group administrators for their input via the Questions section on LinkedIn.

4. **Research LinkedIn People You Know:** Go to the upper right hand box and search the company you are interested in working with. Now go to that company, and you will see a list of your first and second degree connections that work there. You can use LinkedIn to get introduced to someone you want to meet. Just be sincere in your request, and do realize that the first degree connection (the gatekeeper) may or may not forward your request to the person you want to meet.

5. **LinkedIn Direct Ads:** You can advertise on LinkedIn, which can drive prospects to your landing page, group or other destination. You bid on how much you will pay every time someone clicks on the ad. If you pay $1 for each time someone on LinkedIn clicks-through, and it takes 50 clicks before you convert a customer, then you've just spent $50 for one customer. Not bad, if you sell your product or service for $500.

Another cool feature on LinkedIn is LinkedIn Today, which is a news aggregation service. You will find the top stories selected for you based on your industry. Also, check out the Skills section and the Updates Section.

The Skills section lets you evaluate how your skill ranks among the LinkedIn community (whether that skill is growing, and other like skills and how they are growing). This is an excellent place to research how you might expand your business by looking at related skills and adding products and services that are ranked by relative growth and are related to your industry. Adding skills to your profile page is now a must to reach 100% completion on your profile page.

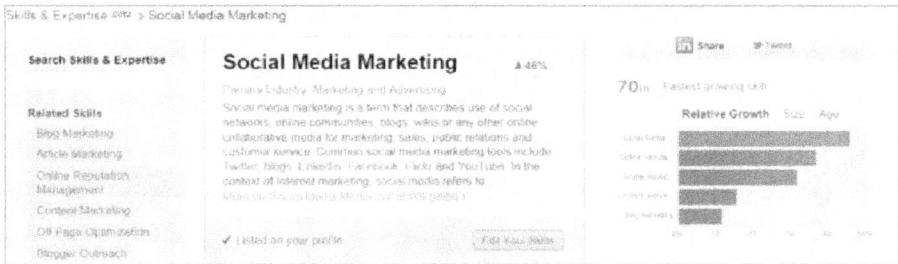

Just remember on LinkedIn, it is the professional's network and you want to put your best face forward... That includes a professional headshot and sharing information that provides value to your connections, groups and other community members.

Twitter to Spice Things Up

Twitter describes itself as "a service for friends, family, and co-workers to communicate and stay connected through the exchange of quick, frequent answers to one simple question: What are you doing?"

If you're new to Twitter, then that description might seem a bit vague and ambiguous. To help you wrap your mind around this short-form messaging tool, start thinking about Twitter as a new form of online communication. Twitter is just communication in a new shape, but it's also a platform for listening to the communication of others in new ways.

Twitter is a micro blog (www.twitter.com). You have 140 characters to use for the messages (tweets) you send. The messages go to your "followers" and in turn, you will receive messages from those you "follow." You have a very limited space to share information about yourself, so use the background area wisely. Include your website link and a short "bumper sticker" bio about what you do. Your background graphic can be customized, or you can use one of the many templates out there. Twitter is a fantastic communication platform and real-time search tool. Here is an example of my custom profile background on Twitter:

Your intention on Twitter is important to your digital body language. This is a platform where you want to "listen" to the conversations occurring before you jump right in. There are certain rules to follow which we will cover here. The best rule is to observe first before you send your first tweet.

Here are some tips for using Twitter effectively adapted from Kerry O'Malley with Marketects (www.MarketectsInc.com):

1. If someone follows you, follow back, unless they are spammers. I always look at someone's Twitter page before I follow back. There are a lot of self-promoters out there, and for me, it's about strategic relationships and not just numbers.

2. A good rule of thumb is to promote others and share value 12 times to 1 self-promotional tweet.

3. Retweet (RT) information that you find valuable, and always give credit to who you got it from by using the @ symbol in front or their handle (name on Twitter).

4. Many people use auto-tweets. I think they are less caring and human, so I am not a big fan of what this digital body language says about someone.

5. Do not self-promote or spam. You will be unfollowed. Use Twitter as it was intended, as a personalized communication tool, not as a way of selling.

Let's discuss Twitter as a marketing tool and some of the rules to follow. There are generally three main categories of users on Twitter. "Relationship Builders" use Twitter as a communication tool, "Promoters" self-promote via this micro blog and "Power Users" have tens of thousands of followers and are constantly tweeting.

These three types of users may not see eye-to-eye on how Twitter should be used. You may view these types as self-serving and only out to make a profit or as arrogant blowhards. It's crucial for you to be who you are as it is a very transparent communication network.

Here is a bit more of a description on the digital body language you might see on Twitter based on the categories described above.

Relationship Builders almost always follow people who follow them. They try hard to insert @ tags which acknowledge others and their tweets. They will send out DMs (direct messages) and expect to get a response.

Promoters rarely post anything of a personal nature. Almost every tweet is about them and what they do for a living or what their company does. They really don't communicate via DM (direct message) unless it brings in a potential customer. They won't use the @ tag unless they are trying to promote their own agenda.

Power Users usually have many more followers than people they follow. They post often. They rarely use the @ tag or DMs. It's all about them. Celebrities or business gurus most often fall into this category.

The bottom line on Twitter is to be very aware of your digital body language and the message you are sending to others. As on all of the social media tools, there is the potential for being misinterpreted and thus misunderstood.

1. **Complete your profile information and use your real name**. I didn't look right or left when I created my "handle" @springboardw, and now I realize this was a mistake. People talk to people on Twitter so use your name. Upload a photo of you, not your dog, not your partner or your company logo. (I like to see you better.)

2. **You should have a similar ratio of people you follow** and those that follow back. If you follow 10 times the number of people than follow you, you may be perceived as a spammer. I always look at this ratio when deciding if I want to follow back.

3. Twitter is a public platform so **if you don't want it public, don't post it!** All posts are permanent and will tag to you. Remember that the search engines like Google pick up the conversations by you.

4. **Don't over self-promote**. Once people follow you, they will know what you do. You do not have to remind them of this. If you self-promote, you will lose followers. Not many people like chest-thumpers.

5. **Use DM (Direct Messages) for ongoing private conversations** with someone. No need to take a discussion out there for everyone to read. Twitter can be very "noisy," and using DM cuts down of some of that noise.

6. **Respond to @ and DMs.** This is just common courtesy. If someone asks you a question or comments to you directly, respond to them in a timely fashion. This will say a lot about who you are being in the digital world and definitely gives an indication of your digital body language. Just as on Facebook and LinkedIn, monitor all comments daily.

7. **What would Miss Manners say?** One of my colleagues always asks, "What would Snoopy do?" Well, in this case, think about basic manners and apply them. Be real, however, be respectful of others, and don't exclude others by voicing opinions that might land you in hot water.

8. **Don't Tweet when inebriated!** This actually applies to all the social media tools. Use your head. You don't want to send messages when you are coming from an emotional place. Remember the saying, "loose lips sink ships". Well, the same applies online. Step away from that computer or phone when partying.

Google+ for Enhanced Likeability

Google+ is the latest social network, and it's part of the largest search engine, namely Google. It is going head-to-head with Facebook. Initially launched in 2011, it encourages users to create Circles (lists) of friends, family, business associates and share stories, discussions and photos. What distinguishes Google+ is the ability to have video conference calls called Hangouts. You can invite multiple people to join you in a video chat, and then record it and post it on YouTube, another Google social media network.

Google+ also lets you create a business page, which encourages businesses to grow their brand and following, and, of course, helps them rank higher in the search engines. It's important to remember that Google owns Google+ and having a Google+ personal and business page is important. Go stake your name claim on both personal and business pages.

Only time will tell if Google+ will catch up to Facebook. At the time of publishing, Google+ had 400 million members. Even though Facebook currently has many more members, the launch of the Google+ business page may just put Google in the driver's seat, considering all of the other applications Google offers its users, including Google docs, YouTube, etc.

Let's explore what Google+ is and why it matters, how to set up a Google+ Business Page and the benefits and best practices for marketing on Google+.

Setting up a Google+ page isn't difficult. First, you will need to set up a Gmail account (free) with Google. Use that account to create your page. You can go to www.google.com/+/business/ to set up your page. Just follow the wizard to get started.

You will select the type of business which may be a local business, product or brand, company/institution/organization/arts/entertainment/sports/ other. After selecting your classification, fill in the basic information.

You will customize your brand's profile by uploading a picture/logo and a tagline – remember KISS: "Keep It Short and Simple" with key words.

I suggest you complete your page and, just like a Facebook page, include as much information about your business and brand as possible before you start promoting it to your Circles on Google+. Keep in mind, once you build it, you must add fresh content, respond to fans' comments, be engaging and remember to ask yourself, "What's In It for Them?"

Note: Google's +1 button is similar to Facebook's Share and Twitter's Retweet. These +1 buttons appear in Google's search results and also can be embedded on other websites as well. They are tied to a destination page address, just like Shares or Tweets are. The bottom line is to use social share buttons to make it easy for visitors to share and like your content.

Another feature of a Google+ page is that you can set up Direct Connection with Google. This means that when someone searches for your Google+ page, you just add the + in front of the page name and it will take you directly to that page. You will first need to install the Google+ Direct Connect code on your business Website to help verify that your Google+ page is the "official" page for your business.

To do this, go to your Google+ page and click on the "Connect Your Website" link under the Get Started Section. You will be taken to a screen that gives you six different Google+ button options. One is to display no button and just install the Google+ Direct Connect code on your website.

Okay, enough technical information. Let's get back to marketing best practices. Some tips for optimizing your business presence on Google+ include sharing LOTS of photos. People on Google+ share lots of individual images. Think about which images or slides you can share to start conversations and inspire people to share your content.

Another marketing tool is to add your recommended links and your other social media properties such as your Facebook Business Page. This is located under the About tab of your page. Use this space to share articles, offers and other items that will drive business. Here is an example of my Google+ Business Page:

Don't forget to add your Google+ URL to other places where you include social media links. This can be on all social media sites, your email signature, email newsletter, on your blog and on your websites. Anywhere you have visitors is key to promoting your Google+ page and building a strong Google+ community.

I am sure many of you are scratching your head right now and wondering, Why do we need another social network? SEO (search engine optimization) is one of the most powerful tools that Google+ has to offer a business. Google wants this platform to be successful, and since they are the king of search, using Google+ for business and individually is only going to help your business SEO rank grow. Some of the ways this helps your SEO are:

- When you create a page, you are instantly indexed.

- You are creating more backlinks to your website.

- Direct Connect - Google's new way to search online.

Google+ most likely will be adding lots of new tools to make their business pages more valuable in the future. Time will tell if they give Facebook a run for their money, especially on the business side of the house.

Remember, the benefit of using Google+ is Google's strong tie to search. Google is not just any network. It's the platform that directly impacts search results and is reshaping how our search experience will be handled on Google. Keep a close eye on this emerging social network.

Pinterest for Visual Interest

Following the theme of social soup, I would describe Pinterest as the icing on the cake, the sprinkles on the ice cream. It is the fastest-growing social network ever and is a fun, engaging place to hang out and share information via pictures and videos that relate to your business.

Pinterest is a social network that allows users to visually share, curate and discover new interests by "pinning" images or videos to their own or others' pinboards, or as we used to call them, bulletin boards.

Users can either upload images from their computer or pin things they find on the web using the Pinterest bookmarklet, Pin It button or just a URL. One of the great features of Pinterest is the ease of adding pictures or videos via the pin it button.

You can be on any page on the Web, and if you see a picture or infographic you like, you can click the pin it button and boom, it's that easy to pin it to your board. I use it to pin interesting material I find on social media.

www.pinterest.com/shelleyroth is my Pinterest page. You can also set up a board with your company name as well. If you remember the old cork boards we would use to pin pictures, recipes, To Do lists and anything else using push pins or - going way back - thumb tacks... Well, that's basically what Pinterest is. It is an electronic version of the cork bulletin board, and you can share all of your "pins" virtually with people who follow your boards and boards that you follow. In addition to pictures and infographics, you can also share videos on Pinterest. You can even offer your products or services for sale by adding a price to what you pin. It will put that price as a banner over the item. Here is an example of my first book sale on Pinterest:

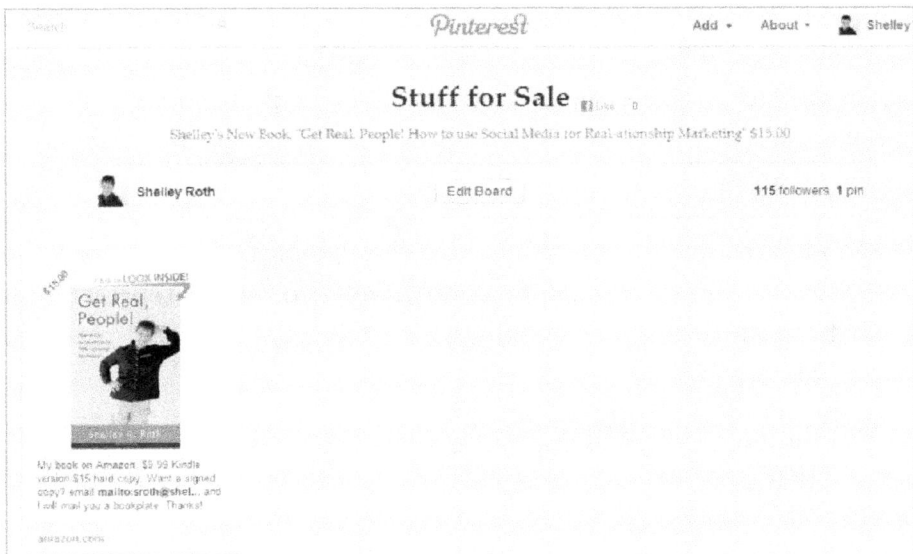

Wikipedia's definition of Pinterest is "a pinboard-style social photo sharing website that allows users to create and manage theme-based image collections such as events, interests, hobbies, and more." Users can browse other pinboards for inspiration, "re-pin" images to their own collections or "like" photos.

Pinterest's mission is to "connect everyone in the world through the 'things' they find interesting via a global platform of inspiration and idea sharing." Founded by Ben Silbermann, Paul Sciarra and Evan Sharp, the site is managed by Cold Brew Labs and funded by a small group of entrepreneurs and inventors.

If your business relies on driving a high volume of website traffic to increase sales, you might consider joining Pinterest. In fact, early research indicates that Pinterest is more effective at driving traffic compared to other social media sites, even Facebook. I also am a big believer in signing up for Pinterest if for no other reason than to stake your claim to your name and help your SEO.

Pinterest realizes the importance of connecting to Twitter and Facebook and therefore enables users to log in using these accounts. Users can automatically post new pins to their Facebook feed for others to see via the Pinterest app. This means more eyes from other channels get access to your pictures.

When Pinterest members are browsing through pins, they also have the ability to share posts through Facebook, Twitter or email. This feature is great to boost social sharing and get your followers to spread the word about your brand.

As you can see below, you can set up as many boards as you like, and pin information you find useful. People can follow what you pin, and you can do the same. My boards are on YouTube, LinkedIn, Facebook, Google+ and also Animals I Love... Don't forget to bring your personality into this and not make it all about business.

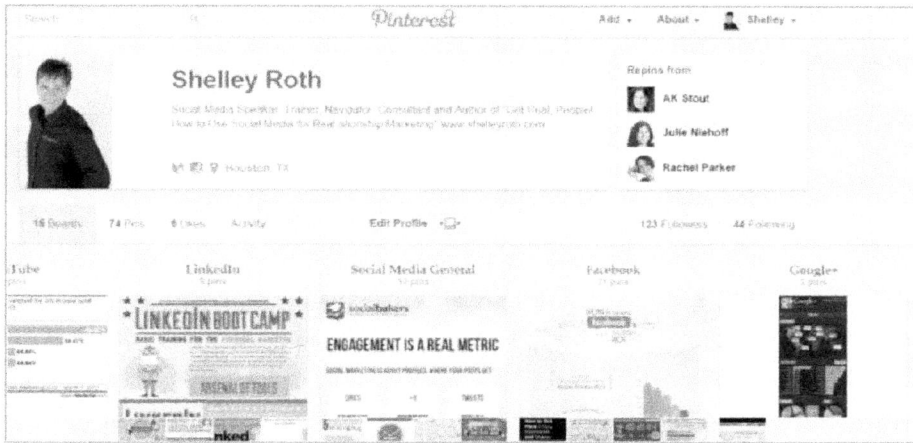

NOTE: Don't forget to read Pinterest's terms of service for copyright and trademark details (pinterest.com/about/terms).

The Secret Ingredient REVEALED

All these great technical tools make for a great social media soup recipe. But **YOU** are the secret ingredient for a successful social media marketing presence. You are the secret sauce and the engaging, REAL-ationships you bring and grow are what deliver the success you will have using all social media tools.

Social media levels the playing field for all businesses, big and small. Its' the engagement that you bring that makes the difference, and it is different for everyone. There is no secret to this. One size doesn't fit all. You must test the market and see what works for your personality and business personality.

Your digital body language (discussed at length in my last book *GET REAL, People!*) is what draws people to you or not. Who you are, the real you, shines through, and it's not just the great engaging picture you post on your social sites.

There are many factors that will determine your success on social media such as your responsiveness to others' comments and questions, your neutrality or lack thereof, your beliefs, likes and interests. All of these contribute to your digital body language and how you are "seen" online. It's the fastest way to make a first impression and a lasting impression. How you respond, what you share, what and who you like, groups you are in, pictures you post, videos you share. All must be considered part of your digital body language and the secret sauce you bring.

The end result is that being aware of "you" as the secret ingredient will help your business not only grow, but thrive. You can have the best content in the world, the best articles, videos, etc., but if you are not sharing you and your true self and engaging with your audience and letting them know the real you, you will not contribute to your businesses growth.

Throwing the old business rule of never making it personal out the window, let contacts know a bit about what you're doing behind the scenes, share news and tidbits with them, and, most of all, be yourself. Just remember: Stay as neutral as possible. An example of what **not** to do was when I shared my passion for the football team I follow with my Facebook fans. I had more people unlike my Facebook Business Page that day than ever! So, leave sports, religion, politics, etc. out of your social presence if you are using social media for business.

Traditional marketing is all about casting a wide net and seeing who you can "capture." Engagement marketing is all about providing value and staying top of mind with your existing customers and prospects by providing what they deem valuable. Engagement marketing is all about building relationships with people that are valued. This is accomplished by making plenty of deposits in their bank before you need to ask for a withdrawal, by giving to grow vs. giving to get.

Sales sharks and traditional marketing are all about "closing" the deal. Engagement marketing/relationship marketing focuses on a dialogue with your prospects and customers vs. a monologue. In this way you are building relationships that last, based on trust vs. one-off sales that are not ongoing.

People want to be heard so let them know you are listening. People want to feel valued and important, so let them know you appreciate them. And people want to know that you are always looking to improve your brand, product, service as a result of their input, so ask for their feedback. The days of a one-way, push approach are gone. It's all about conversations that build community, engagement and brand recognition, which end in the natural by-product of sales. Those companies that engage with their community will thrive in the long run.

Tips for Interacting with Social Media

If you consider yourself a more introverted person, and you aren't so good at knowing what to say to start a conversation, you can still interact on social media successfully.

Here are some tips:

1. Join your college and/or past company alumni group on **LinkedIn**, if you feel part of those associations, and participate in some of

the polls and discussions occurring. You already have something in common with these group members right from the start. In addition, consider joining special interest groups that represent your skills, and join in the activity in those groups.

2. On **Facebook**, like pages of products and services that you use, and then follow those companies and the social interaction that is occurring around them. If someone makes a comment or asks a question on the page, jump in and share your knowledge and participate in the discussion. (But be careful about selling vs. informing or helping by sharing your insights and expertise! It can be a thin line. To tell the difference, visit the discussions on different pages and note when the types of comments and language used make you feel like someone is selling or overselling or just participating to make a pitch rather than genuinely help.)

3. On **Pinterest**, you can create a board with something you are passionate about like a pet board. You can pin pictures of a certain breed of dog you like or cat you fancy. Others will like or comment, and you will have started a dialogue with other people on Pinterest.

4. If you are a natural teacher, you can use low-cost tools to make valuable videos to post on **YouTube**. Maybe they're about your areas of expertise in business or your interest. This is a great way to Give to Grow by sharing information that others will find valuable.

5. On **Twitter**, you might follow other conversations that are interesting to you, and retweet them. This is a simple way to jump into a conversation without having to start one.

Case Study:
Energy Arts Alliance's Real-Life Social Media Story

Energy Arts Alliance is my energy advisor and also my book editor. Over the course of our editing work together on my books, we've had some great conversations about how this Give to Grow organization has used social media successfully.

Here's their story based on an interview I did with Energy Arts Alliance Co-Founder and Chief Experience Guide, J-Coby Wayne. I've included it to give some real examples of how this organization that was built and operates completely from the "Give to Grow" approach (they were the ones who first taught it to me) uses social media so you can be inspired to use it creatively yourself.

Energy Arts Alliance doesn't use social media to build brand or generate followers. We exist to contribute to the community, raise awareness, evolve world relations, evolve people as citizens and support world progress by catalyzing the growth and contributions of pioneering people and enterprises seeking to live, noble adventurous lives. Our community is interested in doing well by doing good. So we have a very different approach to business and social media in general. We are a cause-based, for-profit enterprise that is completely unlinked from fee-for-service commerce as compared to traditional business. We don't have clients; we have partnerships with what we call "Patrons of the Energy Arts."

We have four major ways in which we use social media to fulfill our purpose and mission in service to the community:

Email marketing campaigns. *We are very targeted and looking to serve a very niche market of people whose primary focus is making a difference through their art or entrepreneurial process or job in their corporation, as a student or as person who is very international and considers themself a world citizen. People who are well-traveled and well-read. Social media hasn't been the easiest thing for us to use because we are not going after the majority or consumer stuff that trends.*

With social media, we found that you should use the tools that click with the vibe of your organization. We tried a lot of different email outreach solutions and settled on the one that we felt was closest in vibe to who we are. The one we chose to use is very funny and playful. I learned this when I was setting up our Energy Arts Action eblast service. I forgot to set the schedule date and time, and it was defaulted to a past date. The mail service provider delivered the following message, "Please set your date for some time in the future 'cuz time travel is really hard." Those are the type of messages they send out... When you get a new subscriber, they send a "Doesn't it feel good to be loved" notification message to us. Our own vibe is playful, light, adventurous, so it's an ideal match. I like that it is an "anti-business suit" email program in look and approach, and the tool has allowed us to evolve our ways of reaching out to the customer in the service offering itself.

In terms of how we've benefited from the email service we use, we have become more streamlined and given our community more ways to interact with us via a daily email opt-in quote and question service. It serves as a daily tip. It helps stimulate people into thinking in new ways by offering an idea and a question to prompt your thinking with how are you going to apply the idea today.

This service doesn't translate into direct dollars. It translates into something we consider much more important: A citizen who is thinking more proactively and contributing with new understandings we've helped provide. Therefore, we are doing our job.

We love Facebook. *Of all the social networks, it's our favorite tool because It's just so darned fun. We could write a whole separate book about our experience with Facebook, so we'll just share one experience that has been most surprising. Being philosophical and kind of bookish ourselves, we love to post high-falutin' ideas and tidbits that we hope will inspire our community. So imagine our surprise when our community – people who have philosophical temperaments and intellectual tendencies themselves – started responding in comparative droves to the photos we were posting of vegan meals we were eating on our various travels around the globe! Did anyone respond much to the inspiring story of kids making a difference or the first female Middle Eastern astronaut in space...Noooooo! They commented on and shared our photos of our disgustingly huge vegan comfort breakfast in Seattle. They loved the photo of Kain, my partner in business and life, standing on the footbridge in front of the space-age looking Pritzker Pavilion in Chicago. They loved our energy arts perspective movie review of "Sucker Punch." From then on, we completely flipped our way of interacting with the community, providing 90% fun, personal stuff largely centered on our travels around the world to bring new energy to different regions and to support businesses and projects we believe are contributing to world progress and only 10% of high-falutin' stuff. We never would have known to do this without Facebook's community-enabled feedback mechanisms.*

LinkedIn Question and Answer. *One of my earliest experiences using social media came out of a LinkedIn workshop with you, Shelley, and that's where you introduced the concept of Q and A on LinkedIn. After attending the workshop, I went and did Q and A like gangbusters. I have since dropped off, mostly because LinkedIn just isn't our vibe. As you mention in this book, it's very buttoned down, and I don't get inspired going there. (I get too reminded of my corporate days and what a fish out of water I was there!) BUT I found using Q and A engaged people in a much more targeted niche way than anything we had ever done before.*

My experience with the Q and A was that I posted an answer to someone out there, and I got a lot of questions back in response to my answer. That was a mind-blowing example of effortless engagement! Our company was brand new, and the people who came back with questions in response to my answer were exactly our customer base.

One of the people I met runs a diversity consulting company, Diversiton, in Ireland. He was a government minister before that and served on various European Union bodies. We ended up interviewing him for our magazine, and this is a person I never would have met had I not used the LinkedIn Q and A feature.

We meet and attract a mostly international crowd on LinkedIn. We have an international and global reach, and all social media has aided in expanding this, which could not have been done before unless we were hopping on a plane.

Another cool thing that came about was that I went and looked for Groups. The obvious ones were alumni groups, and I signed up for all Brown University groups since that's one of my alma maters. One of the Brown LinkedIn groups was a media group.

I ended up connecting with a guy who was a Brown alumnus who graduated 15 years before I did. He has an art gallery in Shanghai, and he posted a call for artists in the Brown Media Group. He was hosting a digital art exhibit in Shanghai, China, so I submitted a piece of artwork, and it was selected for the show. We ended up being friends, and he offered to represent my artwork in China. So that's another incredible and inspiring example of the reach of LinkedIn.

Icing on the cake was that this relationship with the gallery owner translated immediately into a paying client even though I didn't have that intention in relating with him at all. Not long after I met him, Energy Arts Alliance did a webinar called "Power up Your New Year," and he participated in that webinar. We have continued to stay in touch.

Our surprising experience with Twitter. *Twitter has been the biggest surprise. I was not feeling Twitter. Even though I'm tech-savvy, I had no idea how to use It, and I didn't want the pressure of being expected to microblog five times a day, especially since we often travel to remote places where there's no digital connectivity. We ended up having an interesting experience with it, though.*

In 2011, we participated in the Social Good Summit virtually from our Hill Country retreat outside of Austin, Texas. This event was right up our alley. We participated by the free Livestream instead of traveling to New York. I decided to live tweet during the conference. I just was tweeting interesting quotes or data points that people were giving during the talks.

I couldn't believe the number of followers on Twitter we were getting! I was not saying anything original; I was not giving an energy arts perspective. I was simply reporting what other people were saying, like a man on the street.

I think we got so many followers that day because I was constant in tweeting, the things I was pulling out were very relevant to our audience and the social media summit (and, by extension, its participants) were so enlightened and aligned to our mission. I used Twitter's recommendations on who to follow that were also attending the Social Good Summit. I didn't even know about that feature before that day!

Twitter provided me with a massive list of people who were attending and pointed out those people Twitter thought I might be interested in following based on being part of the Social Good Summit. A lot of the people I followed followed me back. And they were from all over the world, which was very cool. They were resonating with doing well by doing good.

So, for us, clearly the man on the street reporting was extremely effective. I think this is a great approach if you are naturally good at aggregating and filtering data, which I am good at and enjoy very much. (It's how I made my fame and fortune in the Internet world back in the 1990s.) What I haven't figured out on Twitter yet is how to continue to engage that big set of followers we got from the Social Good Summit effectively. I need to figure out how to keep it going. I will look for other events like this. I also will use it when we are on the road – a "Where is Energy Arts Alliance?" series that shares our travel expeditions.

For Energy Arts Alliance, our experience with social media has proven that when you are a better member of the community coming from a place of Give to Grow, your business grows naturally. And, believe me, I was one of the most effective, hyper-competitive sales sharks out there before I got on my path of waking up and becoming a better person focused on serving others and supporting their success.

Is Your "Digital Body Language" Portraying You as a Sales Shark?

Is your "Digital Body Language" portraying you as a Sales Shark?

In my first book, *GET REAL, People!*, we did research via surveys to see what our focus group "reads" from profile pictures and other digital body language indicators. What we learned is this:

When we are online, there are many ways to "read" and "sense" a person's body language, or what I refer to as "digital body language." We immediately determine if we "feel" someone is a sales shark vs. a truly caring, giving person who will bring value to our life. The number one cardinal rule of what NOT to do via social media is SELL. There is no faster way to turn people off in social media land than to pitch and self-promote.

So on the road from sales shark to recovering sales shark, let's start with the **picture** that represents you:

1. **When you put your picture out there, that's the FIRST connection people will make to you.** You want to assure you have an energy vibe that draws people to you. What is an energy vibe? How do you determine that? It really is a gut feeling you get when you are with someone in person. Some people call it an aura, but basically it is the first assumption you create about that person even before they open their mouth. So, that picture you post on LinkedIn, Facebook, Twitter, Google+ and Pinterest is truly worth a thousand words. Make sure it reflects who you are.

2. How you **write/post/respond to other's posts** can tell lots about you. Are you providing what people want to hear? Are you adding value based on what they deem is valuable? Are you engaging with them? I

mean truly engaging vs. what's in it for you? Are you asking for what you want or what they want? Are you giving to get vs. giving to grow? Are you making it easy for your community to respond to you? Do you make it seamless to learn about "brand you?" Do you make it easy to access information or hard? The answers to these questions determine how others perceive you online.

One of my red flags on digital body language is inside of groups. The spirit of groups, whether on Facebook or LinkedIn, is to share and add value to the group. Now that means not promoting your products and services unless they are relevant to the topic at hand. I have left many a group when the intention of the group administrator became using that platform for selling themselves or their products. Most social media tools now have a way of self-policing this issue so the true spirit of the group can occur.

As an example, in group discussions, groups don't want to hear about my workshops coming up. I don't want to share my workshops. It is tempting, but it's not the digital impression that I want to portray out there. It's important to always think before posting anything online.

Now IF the discussion is specific to what you have to offer, well by all means, jump in and share what you can offer, whether it's your product or service or someone you recommend. The key is IF it's **relevant** to the conversation and it adds value.

Our conclusion after this exercise of surveying people on their perceptions of digital body language was that you're never going to know how others may be interpreting you unless you ask yourself these questions first:

1. How am I being interpreted by others?
2. How are my picture and words being interpreted by others?
3. How are my grammar and punctuation?

4. Am I responding in a timely fashion?

It was obvious from the many interviews and surveys we conducted that people's body language can be interpreted online. Whether through a smile, the written word, pictures that are posted, spelling or grammar, responses to comments that are made and conversations we engage in, we all formulate opinions of each other from digital input.

The key to interpretation is to be aware of what your senses tell you when dealing in the digital world, and always consider how you will be interpreted online.

The bottom line is to be conscious of your digital body language. It doesn't hurt to ask people what their impression is of you. Don't be afraid to ask. If you don't want to know, you don't want to grow!

Since we are all working to build a community of customers, prospects, colleagues and partners both online and in life, I want to help you be aware of your digital body language and how you are interpreted online.

Can you imagine a world where we totally trust each other's realness online and offline? We would have more rewards than imaginable. So, the goal is NOT to be selling anything. It will always be about giving to your colleagues, prospects, partners and clients while knowing that when they need what you have to offer, they will then engage your professional services or purchase your products. If you come from this place, you will never have to toot your own horn, "take no prisoners" or close a deal because it's the end of the quarter and numbers must be made.

Let's look at some examples of the obvious and easiest way to read digital body language, the picture you post on Facebook, LinkedIn and other social sites that require this.

Here are some pictures to look at.

Provocative...Not such a good idea to reach your business audience

Photograph: Louisa Stokes, freedigitalphotos.net

Angry and threatening... Is that how you want people to see you?

Photograph: Ian Kahn, freedigitalphotos.net

Happy and light-filled... That's more like it!

Photograph: Dundee Photographics, freedigitalphotos.net

Partying...Remember HR and college admissions offices research prospective employees and students via all social media!

Photograph: photostock, freedigitalphotos.net

Ask yourself: What do you get from the pictures shown? What engages you? Who do you resonate with? And why do you feel drawn to them or not drawn to them? What would you do differently? Why?

The next step in "reading" someone is via the written word. If we didn't have the benefit of seeing pictures, I believe that with a lot of practice, we could read people's bios, blogs and status updates and determine when someone is trustworthy, giving, forthright and honest vs. selfish, self-promoting, self-centered... My advisors at Energy Arts Alliance (energyartsalliance.com) can already do this really well and help others learn how to do it, too. They created the Digital Muscle-Builder Exercises for my first book to help you hone these skills yourself (www.springboardworks.com/books).

If you don't have a picture posted on LinkedIn or Facebook, many people will not even consider responding to you. Be sensitive to how you may be interpreted and what the lack of a picture may say about you.

Personally, I tend to interpret the absence of a photo as meaning that you weren't thoughtful enough to learn how to upload a picture or get a picture that represents you prior to publishing your page or bio. That is purely my perception and might not be reality. The real reason you don't have a photo up may be that you're concerned about privacy issues due to past history and don't want to be exposed online with a photo. If that's the case, then maybe social media isn't for you.

The presence or lack of a photo speaks volumes about a person without saying a word – or maybe causes you to be "saying" something you didn't intend based on someone else's interpretation! Without in-person cues, the potential for these variations in interpretation becomes greater and the gaps between intention and perception can get wider, leading to potential misunderstandings and unintended consequences. Always consider how you will be interpreted.

Shedding the Sales Shark:
Sharing Your Heart & the Real You Within

Thanks for being adventurous to pick up this book and explore shedding your sales shark.

To wrap this adventure up in the spirit of giving to grow, I want to thank some special people who have helped and supported me in my continuing transformation to be my real self on this planet we call Earth.

Two of these special people are my energy advisors, J-Coby Wayne and Kain Sanderson, and their organization Energy Arts Alliance. Without their amazing guidance and direction, this book would not be. They were the ones who first exposed me to two core philosophies of this book: Give to Grow and the Law of Abundant Exchange. They live these philosophies in everything they do, and as a result, they attract many opportunities to help people like me in really profound ways. Their support, vision and belief in my higher purpose are why this book exists. This is the second book they have guided me through.

A few other people crucial to this book are Yemi Owolabi, the photographer, Tric Ortiz, the graphic designer, and Maya James, the make-up artist who did the make-up for the cover photos. Another special person I want to thank is my partner, Cara Hawthorne. She always has my back.

I also thank all of my many fans, connections, followers and subscribers who inspire me to Give to Grow every day of the year. Finally, thanks to my colleagues who shared their stories with me about leaving corporate jobs behind and venturing out on their own.

What I hope you have received from this book is the opening to not question your heart and just being who you are to pursue your passion and purpose. If you have identified something that has been a seed or spark, and you acknowledge it and nurture it, then this book has served its purpose. Living in the preconceived notions of who we are is not truly serving the world. Living life as someone you are not is not serving humanity to the best of your ability.

Social media is a great platform to leave the shark tank, come out of the closet and be the person you are intended to be, sharing your real self with the rest of the world and watching your business thrive through the amazing tools at our disposal via social media networks.

See you online! You can find me there at www.shelleyroth.com.

www.ingramcontent.com/pod-product-compliance
Lightning Source LLC
Chambersburg PA
CBHW071910200326
41519CB00016B/4554